General editor: Andrew S. Thompson
Founding editor: John M. MacKenzie

When the 'Studies in Imperialism' series was founded by Professor John M. MacKenzie more than thirty years ago, emphasis was laid upon the conviction that 'imperialism as a cultural phenomenon had as significant an effect on the dominant as on the subordinate societies'. With well over a hundred titles now published, this remains the prime concern of the series. Cross-disciplinary work has indeed appeared covering the full spectrum of cultural phenomena, as well as examining aspects of gender and sex, frontiers and law, science and the environment, language and literature, migration and patriotic societies, and much else. Moreover, the series has always wished to present comparative work on European and American imperialism, and particularly welcomes the submission of books in these areas. The fascination with imperialism, in all its aspects, shows no sign of abating, and this series will continue to lead the way in encouraging the widest possible range of studies in the field. 'Studies in Imperialism' is fully organic in its development, always seeking to be at the cutting edge, responding to the latest interests of scholars and the needs of this ever-expanding area of scholarship.

Engines for empire

Manchester University Press

SELECTED TITLES AVAILABLE IN THE SERIES

WRITING IMPERIAL HISTORIES
ed. Andrew S. Thompson

MUSEUMS AND EMPIRE
Natural history, human cultures and colonial identities
John M. MacKenzie

MISSIONARY FAMILIES
Race, gender and generation on the spiritual frontier
Emily J. Manktelow

THE COLONISATION OF TIME
Ritual, routine and resistance in the British Empire
Giordano Nanni

BRITISH CULTURE AND THE END OF EMPIRE
ed. Stuart Ward

SCIENCE, RACE RELATIONS AND RESISTANCE
Britain, 1870–1914
Douglas A. Lorimer

GENTEEL WOMEN
Empire and domestic material culture, 1840–1910
Dianne Lawrence

EUROPEAN EMPIRES AND THE PEOPLE
Popular responses to imperialism in France, Britain, the Netherlands, Belgium, Germany and Italy
ed. John M. MacKenzie

SCIENCE AND SOCIETY IN SOUTHERN AFRICA
ed. Saul Dubow

Engines for empire

The Victorian army and its use of railways

Edward M. Spiers

MANCHESTER
UNIVERSITY PRESS

Published by MANCHESTER UNIVERSITY PRESS
ALTRINCHAM STREET, MANCHESTER M1 7JA
www.manchesteruniversitypress.co.uk

British Library Cataloguing-in-Publication Data
A catalogue record for this book is available from the British Library

Library of Congress Cataloging-in-Publication Data applied for

ISBN 978 0 7190 8615 1 hardback

First published 2015

Typeset in 10/12pt Trump Mediaeval
by Graphicraft Limited, Hong Kong
Printed in Great Britain
by TJ International Ltd, Padstow

CONTENTS

MAPS AND ILLUSTRATIONS

Maps

Illustrations

ACKNOWLEDGEMENTS

Quotations and references from the Royal Archives appear by the gracious permission of Her Majesty Queen Elizabeth II.

Quotations from the Londonderry correspondence appear by the permission of Lord Londonderry and the Durham County Record Office; quotations from the Haldane papers appear by the permission of The National Library of Scotland; and the reference to the Tiplady diary is by permission of the Blackburn with Darwen Library & Information Service. Quotations from the papers of General Archibald Hunter appear by permission of Nicol John Hunter Russel; from the diaries of Henry Wilson by permission of the Trustees of the Imperial War Museum; from the papers of J. F. Maurice by permission of The Trustees of the Liddell Hart Centre for Military Archives; from the diary of Captain Fitzgibbon Cox with the permission of Lincolnshire Archives and the Trustees of the Royal Lincolnshire Regiment Museum; from the papers of the third marquess of Salisbury by Sarah Whale (Archives Department, Hatfield House); and from the Wingate diary by permission of Martin Dane.

I am also grateful for the assistance of Miss Pamela Clark (Senior Archivist, Royal Archives); Mrs Lynette Beech (Royal Archives); Dr Alastair Massie (National Army Museum); Roderick McKenzie (Regimental Headquarters, Argyll and Sutherland Highlanders); Lauren Jones, Adam Walsh and Danielle Sellers (Royal Engineers Museum, Library and Archive); Ms Sara Basquill (Collections Access Officer, Museum of Lincolnshire Life); Andrew Wallis (Guards Museum); Nicola Wood (Archives Assistant, Queen Mary, University of London), Jackie Brown (British Library); Dr Antonia Moon (curator of post-1858 India Office Records, British Library); Dr Maria Castrillo (curator of Political Archives, National Library of Scotland); Lianne Smith (Archives Services Manager, Liddell Hart Centre for Military Archives, King's College Library, London); Vicki Perry (Head of Archives and Historic Collections, Hatfield House); Jesper Ericsson (Gordon Highlanders Museum); Liz Bregazzi (County Archivist, Durham County Record Office); Mary Painter (Librarian Community History and ICT Customer Services, Blackburn Central Library); Ms Jane R. Hogan (Senior Assistant Keeper, Sudan Archive Durham University); and Anthony Richards (Head of Documents and Sound, Imperial War Museum). I appreciate, too, the assistance of the staffs of the British Library (Newspaper Collection) at Colindale, the Special Collections,

Brotherton Library, University of Leeds, the Templer Study Centre of the National Army Museum, the National Library of Scotland and the inter-library loans service of the University of Leeds.

I am also most grateful to Omar Khan for permission to reproduce photographs from his collection, to Peter Harrington for the provision of several images, and for the permission to reproduce them, from the Anne S. K. Brown Military History Collection, Brown University, Rhode Island, and to Danielle Sellers for permission to reproduce an image from the collection of the Royal Engineers Museum, Library and Archive.

I appreciate, too, the support from various academic colleagues, namely Professors Graham Loud and Malcolm Chase (Leeds University), Dr Jeremy Crang (University of Edinburgh), Professor William J. Philpott (King's College, London), Professor Andrew S. Thompson (University of Exeter), and, for her encouraging words, Dr Rachel Utley (Leeds University). I am particularly grateful to David Appleyard for his preparation of the maps for this volume.

As ever I am indebted to the patience and forbearance of Fiona, my wife, Robert and Amanda as they endured the preparation of another book.

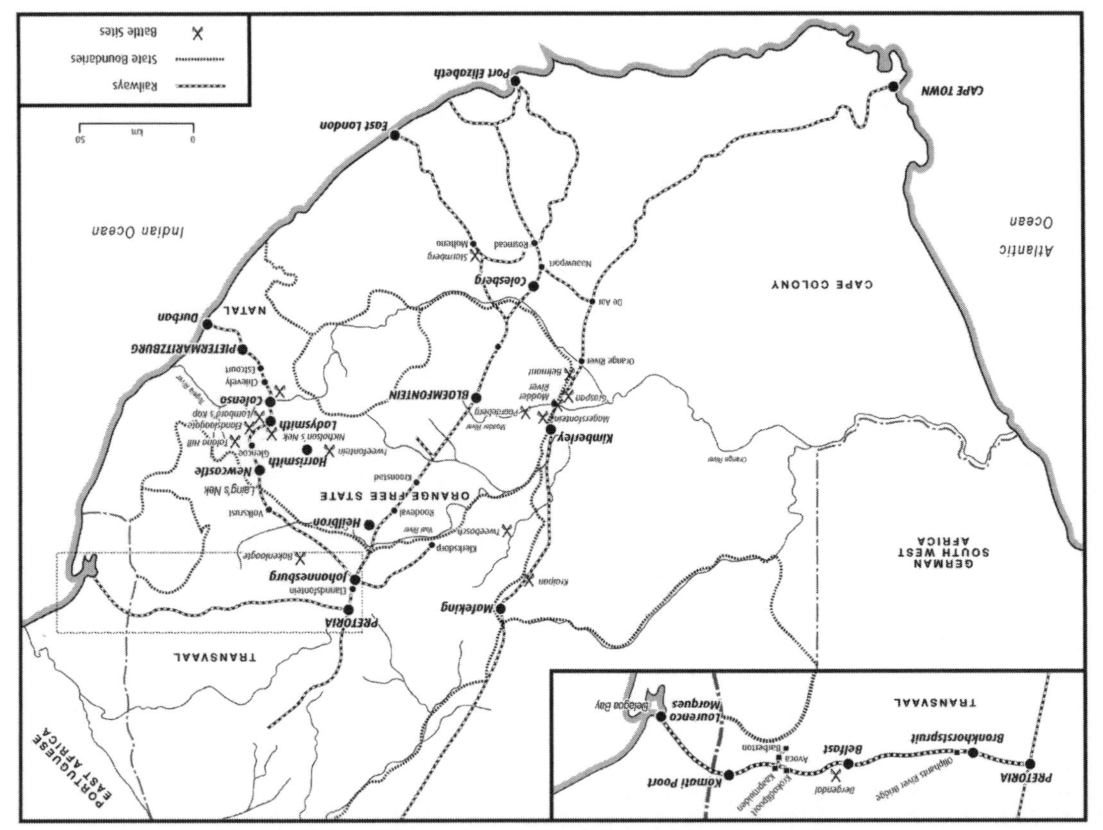

Map 1 South African railways

1

Public order: the army and railways

The railway represented one of pivotal technological developments of the nineteenth century. In Britain, wrote W. T. Jackman, the appearance of the railway had a 'grandeur and ostentation that charmed the public. It seemed the embodiment of enterprise and boundless capabilities.'[1] For many Victorians, and railway historians, the primary benefits of the railways were socio-economic: they accelerated the movement of goods and people, connected disparate communities and facilitated the transmission of news, images and information.[2] Similar benefits would flow across the empire, opening up the hinterland in Canada, India and South Africa, and enhancing the development of commerce, free trade and prosperity. As the 'largest single investment of the age',[3] they overcame barriers of time and space, had a huge impact upon the economy of India, and enhanced Victorian understanding of the empire through the transmission of images of peoples, places and events.[4]

The onset of the 'railway age',[5] with the Stockton and Darlington line opening for colliery traffic in 1825 and the Liverpool and Manchester line for passenger and goods traffic in 1830, coincided with recurrent outbreaks of public disorder. Whether these events occurred in urban or rural communities, and whether they were triggered by economic discontent, localized agitation, radical demagoguery or a combination of all three, they often involved actual or potential threats to property. As the local magistrates often reported such events in a state of panic, the state had to respond but, in so doing, could not exploit the potential of railways until private investors had laid the critical lines. Fortunately, the astonishing example of George Stephenson's 'Rocket' in the Rainhill Trials of 1829, where it managed speeds of nearly 30 mph (48.3 km per hour), and then the example of a single company operating the Liverpool and Manchester Railway, inspired the first railway 'boom' when expenditure on railways increased from £1 million in 1834 to £9 million in 1839. The key lines were built, particularly the Grand Junction and the London and Birmingham lines in 1837 and 1838 respectively, which linked the capital by rail with the industrial towns and ports of Lancashire. Thereafter amid the railway-building 'mania' of 1845–49, peaking in 1847, the railway network expanded from 2,409 km in 1840 to 9,791 km in 1850, linking the major population centres of England, Scotland and Wales.[6]

That railways would come to assume a considerable significance in the maintenance of public order reflected the continuing role of the army in providing aid to the civil power. This role persisted despite the ratifications of the Metropolitan Police Act (1829), the Lighting and Watching Act (1833) and the Municipal Corporations Act (1835). Although the last Act required reformed corporations to establish watch committees, which could then appoint constables to be paid at the expense of the ratepayers, many corporations were reluctant to do so. As late as 1849, there were at least twenty-one corporate towns (or 12 per cent of the total number) that had not established a police force. Even more boroughs were reluctant to establish a sufficiently large police force, and so they complied with the letter if not the spirit of the legislation, and established police forces that were woefully under strength. Whereas the Metropolitan Police had established a ratio of constables to citizens of 1:443 by 1840, only 1 in 20 boroughs attained ratios of 1:600 or better from 1839 to 1848, and barely one-quarter of the provincial boroughs maintained a ratio of 1:1,100 throughout the period of the Chartist disturbances (1837–48).[7]

More recent research has challenged the traditional Whig narrative that Britain was a largely unpoliced society prior to the 1829 Act, and has shown that there was private policing in parts of London before the Act, and that a policing system existed at parish level, albeit one limited in scope.[8] There were also pockets of experimentation in policing, including a Cheshire Police Act passed in 1829 and private Policing Acts in the burghs of Scotland.[9] Yet the absence of regular police forces was manifest in many of the northern industrial communities of Yorkshire and Lancashire, where the People's Charter had an early appeal in 1838.[10] Advocating radical political reform, orators attracted noisy meetings and led parades, many of them held by torchlight at night, involving thousands of people and inducing widespread alarm among the property-owning classes. In Manchester, where a police 'force' of two constables and seventy-four watchmen had proved utterly ineffectual in the face of industrial disturbances and riots in 1829, the police, in 1837, amounted to only 30 constables, 150 watchmen and several hundred special constables.[11]

Accordingly, the state, county and municipal authorities looked to the military (both regular and auxiliary forces), part-time special constables and later military pensioners to provide aid to the civil power. This was nothing new for the regular army: even at the height of the Peninsular War, in 1812, more than 12,000 soldiers, including militiamen and yeomanry, were deployed between Leicester and York to suppress the Luddite disturbances.[12] The military provided all manner of assistance. In coastal communities, particularly in Cornwall, the military aided customs and excise officers in the seizure of contraband and in countering smugglers. In Ireland military units provided escorts for prisoners and witnesses, guards at gaols and executions, protection for sheriffs, bailiffs and excise officers in their periodic attempts to curtail the distilling of illicit whiskey and a presence at public gatherings, such as fairs, markets and political meetings, where breaches of the peace might occur. They were deployed extensively during elections, acting as escorts for voters and poll books and serving as a riot-control force if necessary. Above all, in the early 1830s the military, including the Irish yeomanry until their abolition in 1834, assisted the Irish Constabulary during the tithe war,[13] when magistrates sought to enforce the collection of tithes on behalf of the Church of Ireland or the seizure of goods in default of payment. There were several bloody confrontations during this 'war', when the police and military fired on mobs, killing and wounding protestors and sometimes suffering fatalities themselves.[14]

However experienced in these multifarious duties, army commanders knew that these events were always risky and unpredictable. In the aftermath of Waterloo (18 June 1815), they were made even more demanding by the rapid and extensive cuts in military expenditure and manpower made by successive governments. As the state rushed to reduce the military-fiscal burden of wartime, it cut expenditure on the army and ordnance from £43,256,260 in 1815 to £10,699,865 in 1820, and thence to below £10 million in the 1820s and just under £8 million by 1836. It allowed this expenditure to rise only during the Chartist disturbances, and subsequent war scares, to reach £9,635,709 in 1853. Military numbers fell in line with the financial cuts as the army slumped from 233,952 men in 1815 to 114,513 in 1820, and to 104, 066 in 1830, before rising slightly to 124,659 in 1840 and 136,932 by 1850. Even worse, the garrisoning of the empire consumed at least half of the army, leaving only 64,426 officers and men in the United Kingdom in 1820 and a mere 44,731 by 1825. With the Guards normally based in London,[15] the other home-based infantry and cavalry units, scattered across the country, were frequently on the move. The 1st Royal Dragoons, for example, moved from Lancashire in 1820 to Dorset in 1821, and then, on half-yearly rotas, to Kent, London, York, Edinburgh, Dundalk, Dublin, Newbridge, Cork and Ballincollig, before returning to Lancashire in 1829.[16]

Hampered by the shortage of regular soldiers, the state was unable to compensate by drawing upon substantial numbers of auxiliary forces.

During the French Revolutionary and Napoleonic Wars, it had enrolled large bodies of militia, mounted yeomanry and volunteers primarily for home defence. All these forces, though, had aided the civil power, suppressing food riots, Luddite disturbances and riots or mutinies among the local militias. By 1808, most of the volunteers were incorporated into the semi-balloted local militias, and all of the local militias were disembodied in 1816. The militia, too, were disembodied apart from their permanent staff at the end of the Napoleonic Wars.[17] Only the yeomanry survived but in a severely truncated form, with much of their strength (17,818 men in 1817) concentrated in the home counties, East Anglia, the Midlands and the maritime counties of southern England.[18] More expensive than the regular forces (as they had to be paid for their voluntary services whenever called out in aid of the civil power), the yeomanry were also much less popular after the events at St Peter's Field, Manchester, on 16 August 1819, popularly known as the Peterloo Massacre. Faced with a vast crowd of possibly 60,000, sixty cavalrymen of the Manchester and Salford Yeomanry were ordered to assist in serving an arrest warrant on Henry Hunt and other radical orators. They became trapped, and escaped only with the assistance of the 15th Hussars, but during the resulting mêlée eleven people died and another 400 were injured (more by crushing than by sabring),[19] leaving the yeomanry's reputation in tatters. The Manchester and Salford Yeomanry was disbanded in 1824, and the government briefly tried to disembody the entire force in 1827–28 before an upsurge of disturbances in rural areas (the Swing riots of 1830–31) occasioned the restoration and renewed use of the yeomanry corps. Although Lord John Russell, when the Whig home secretary, declared that 'he would rather that any force should be employed in case of local disturbances than the local corps of yeomanry',[20] and distaste for employing the costly and unpopular yeomanry persisted, both Tory and Whig governments had to employ their services extensively during the 1840s.

Accordingly the army, sometimes bolstered by the support of ex-army pensioners, remained the principal military instrument in aid of civil power. When they served in this capacity, soldiers acted under the control of the civic authorities, with the home secretary assuming responsibility for the distribution of troops across the United Kingdom, though usually after consultation with the Horse Guards and the commanders of the military districts. At local level, the local magistracy had the responsibility for maintaining public order. In the event of a public disturbance, they had to gather sufficient forces, relying upon local police in the first instance, but if two or more magistrates were present, they could swear in special constables. They could also request aid from the local military or call out, on their own authority, the local yeomanry. They then had to lead this force to the scene of the disturbance, decide whether to read the Riot Act (whereupon the riot became a felony and the authorities could use force, including firearms, to suppress it) and give the order to open fire.[21]

This process could prove disastrous. On 2 June 1831 an angry crowd of 2,000 protestors, demanding reductions in the price of bread and an increase in wages, assembled outside the Castle Inn, Merthyr Tydfil. Inside the inn, local employers and magistrates were meeting Richard Hoare Jenkins, the High Sheriff of Glamorgan; they rejected the demands of the mob, which not only refused to disperse but also then attacked the inn. Jenkins panicked and read the Riot Act before a small detachment from the Reserve of the 93rd (Sutherland) Highlanders were properly deployed in support of the special constables. In the resulting confrontation, six of the sixty-three Highlanders including their commander, Major Thomas Falls, were badly injured, at least sixteen people died, and another seventy were wounded. Compelled to withdraw from the inn to the more defensible Penydarren House, the magistrates and military abandoned the town for eight days as the rioters commandeered arms and explosives, set up road blocks and attacked the military reinforcements. They ambushed the 93rd's baggage-train under escort of forty of the Glamorgan Yeomanry; humiliated the Swansea Yeomanry by disarming them in an ambush and throwing them back in disorder to Neath; beat off a relief force of a hundred cavalry sent from Penydarren House; and organized a mass demonstration against Penydarren House. Only after the arrival of another 450 soldiers were the authorities able to regain control of the town.[22]

Among the Reform Bill riots of the same year, mob rule prevailed again when the military withdrew from the centre of Bristol. Under the command of Lieutenant-Colonel Thomas Brereton, two troops of the 14th Light Dragoons and a troop of the 3rd Dragoon Guards had arrived in the city on 29 October, when rioters were already attacking the Mansion House. Lacking any orders to fire, Brereton withdrew his ninety-three dragoons, allowing a mob that would eventually number several thousands to pillage and burn the principal buildings. Over three days the rioting continued until the mayor, Charles Pinney, authorized Brereton to act on 31 October. Although Brereton still dithered, Major Mackworth gave the orders to attack and the dragoons swept across Queen Square, clearing the rioters and inflicting over 100 casualties. Writing about these events in his diary, Edward Law, the first earl of Ellenborough, reflected upon the extensive destruction of property, including 'the Bishop's Palace, the Custom House, the

Mansion House and the three prisons': he dolefully observed, 'I fear there are very few Troops at Bristol.'[23] Compounding this disaster, Brereton was later court-martialled for leniency and, on the fourth day of his trial, shot himself.

If these were among the more serious disturbances, the Reform Bill riots had occurred in many localities (notably Derby, Nottingham and Mansfield in the Midlands and, on a smaller scale, Exeter, Yeovil and Blandford in the West Country), indicating that the maintenance of order could stretch the resources across the country. Accordingly, both the state and the army soon saw that the new network of railways offered a potential means of responding to challenges in different parts of the United Kingdom, and of doing so with relative alacrity. Ironically Britain's most famous general, Arthur Wellesley, the first duke of Wellington, had experienced the speed and power of the railways at first hand during the opening of the Liverpool and Manchester Railway on 15 September 1830. Wellington, who served twice in this period as commander-in-chief of the army (in 1827–28 and 1842–52) and twice as prime minister (in 1828–30 and 1834), was then prime minister. He led an unpopular Tory government that had fractured its own support over Catholic emancipation while remaining opposed to political reform. The railway company had hoped to mend some of those political fissures by inviting Wellington and many leading Tories, including the marquess of Salisbury, Sir Robert Peel and the leader of the more liberal Tory faction, William Huskisson, the MP for Liverpool, to the opening ceremony. While the duke was hugely impressed by the experience of travelling on a railway coach at speeds of 26 km per hour, and occasionally at 48 km per hour, with trains passing each other on the two lines, the whole event was overshadowed by an accident at Parkside, where the duke's train stopped to take on water. As Huskisson's party descended from their coach to meet the duke, Stephenson's 'Rocket' rushed down the other line, inducing panic in the portly and enfeebled Huskisson, who fell on the line and had his thigh crushed by a wheel. As Wellington subsequently encountered a very hostile mob at Manchester, before learning of Huskisson's agonizing death later that night, the whole experience, as his biographer remarks, 'prejudiced the Duke for ever against railways'.[24]

Nevertheless, the practical utility of railways was all too obvious even at a time when the railway network was far from complete. Soon after its opening in 1830 the Liverpool and Manchester Railway was the first railway to carry soldiers on active service, saving a two-day march from Manchester, and, on 10 July 1832, the 91st (Argyllshire) Regiment had its first experience of travelling by train from Manchester to Liverpool prior to embarking on two steamers to Dublin.[25]

Fortunately, from the perspective of the authorities, they were not challenged too much in the populous industrial region of the north until the onset of significant Chartist disturbances in 1839. Within the Northern District, a military district that encompassed eleven counties in the north of England, and had its headquarters in Manchester, the troop levels fell from 7,280 in 1831 to under 5,000 until 1839, whereupon the numbers rose steadily, peaking at 8,185 in May 1840 before easing down to 5,080 in 1841.[26]

Prior to the movement of substantial numbers of regular soldiers northwards, Metropolitan policemen were despatched regularly to the provinces to make up for the deficiencies of local constabularies. A total of 2,246 policemen were sent out from London between June 1830 and January 1838, an average of some 300 per annum but rising during the anti-Poor Law disturbances to 444 in 1837 and 764 in 1838 respectively. These policemen often received hostile receptions because many deemed them unconstitutional and quasi-military, since they wore uniforms (blue swallow tail coats) and carried wooden truncheons. They also discouraged the local authorities from providing for their own defence, were too small in number to deal with serious riots and had to act under the direction of local magistrates. They often proved less successful in controlling crowds than they were in London partly on account of inadequate numbers, and partly because they had less knowledge of the localities in which they had to operate. Sometimes their presence provoked the mob, notably at the Bull Ring in Birmingham on 4 July 1839, when they charged into a crowd of about a thousand people to arrest a Chartist speaker, precipitating such a violent response that the 4th Dragoons had to rescue the police.[27]

While the practice of sending Metropolitan policemen around the country diminished over the years (as provincial authorities raised their own police forces), the movement of large bodies of soldiers became more prevalent. The railway companies, all privately owned bodies, readily assisted. Some of these companies had close connections with the armed services, employing retired officers as secretaries and general managers, and they developed a corporate culture in which their companies functioned in a disciplined, hierarchical manner with uniformed staff.[28] They found the requisite coaches to convey soldiers and their families across country, buildings in some stations to serve as temporary barracks and work-people to serve as special constables protecting railway property. They also contributed indirectly, by enhancing communications through access to their telegraphic equipment, which ran alongside the lines to improve traffic control.[29]

Faced with anti-Poor Law agitation in the late 1830s, and the opening phase of Chartism in the Northern District, the commanding officer, Major-General Sir Richard Downes Jackson, sought additional men from Ireland, and readily exploited the rail networks. Both he and his successor, Sir Charles James Napier (March 1839–September 1841), benefited from the relative tranquillity in Ireland, which enabled three infantry and three cavalry regiments to be brought over from Ireland. They were landed at Liverpool and moved by rail to Manchester. Of the three units brought over in May 1839, the 1st Royal Dragoons, the 10th (North Lincolnshire) Regiment of Foot and the 79th (Cameron) Highlanders, Napier remarked that 'the last being in kilts terrified the Chartists more than a brigade of other troops'.[30] For movements along the London and Birmingham Railway, a large military depot was established at Weedon in Northamptonshire. The aim of the policy was to bring large concentrations of troops together in critical districts, from which forces could be sent into disturbed areas where necessary. Many of these movements across the districts had to be made by foot or on horseback because apart from four railway lines, including the Manchester and Leeds line, the West Riding still lacked an extensive rail network. In bringing soldiers into Manchester, the heart of the Lancastrian industrial region, Napier sometimes tried to deceive the local agitators by a phased use of the rail network: as he informed the under-secretary of state on 25 May 1839, 'One wing of the 10th came by a morning train yesterday; the other by an evening train, which made everybody suppose two regiments had arrived.'[31]

By the end of 1839, the state had concentrated 10,527 soldiers in the military districts affected by Chartist disorders, with 7,686 men in the Northern District, 969 in the Midland command and 1,872 in South Wales,[32] where the largest armed insurrection of the Chartist era had erupted in Newport on 4 November. The killing of twenty-two people and the wounding of another fifty underlined the risks that could occur when a small company of soldiers (two non-commissioned officers (NCOs) and twenty-eight men of the 45th (Nottinghamshire) Regiment under Lieutenant Basil Gray) and some special constables (most of the 500 'sworn in') faced an all-out assault from a Chartist mob of between four and five thousand, armed with muskets and pikes. Thereafter as fears of further disturbances persisted, almost a thousand soldiers were rushed into South Wales to be billeted in Newport and later in Cardiff.[33]

Rushing large numbers of soldiers into a district by rail, however, posed numerous problems, and these were widely recognized at the time, not least by Charles Napier, arguably the most successful of the district commanders. An extreme radical who detested the new Poor Law and sympathized with the plight of the poor and the political aims of the Chartists, he was a courageous appointment.[34] While

Napier believed that the Whig government should seek a political solution to the rising tide of discontent, he could not countenance direct action: 'Bad laws must be reformed by the concentrated reason of the nation gradually acting on the legislature, not by the pikes of individuals acting on the bodies of the executive.'[35] So he accepted that he had to contain the threat from the Chartists locally but hoped to do so without spilling blood on either side. 'I dread bloodshed', he wrote, wanting to avoid both 'a terrible slaughter of the unhappy Chartists' and any military disaster: 'a military mishap would be a national misfortune'.[36]

Within this context he harboured all manner of anxieties and apprehensions about the security of his soldiers when they were brought into a district where there was a conspicuous lack of suitable barrack accommodation. He complained bitterly about soldiers being scattered in twenty-six detachments across his eleven counties, with some units 322 km from him, and often in 'disgusting' and 'dangerous' quarters, including public and private houses, the worst being the forty-two troopers quartered in twenty-one billets within Halifax.[37] His fears were twofold: first, that soldiers individually or in small groups could be subverted in their loyalty, and that reports of Chartism finding adherents among the Rifles only underscored his preference for relying upon 'troops from Ireland' and 'Irish rather than Scotch, and Scotch rather than English',[38] and second, that soldiers living in improvised barracks could be vulnerable to attack. 'Chartists', he feared, 'may place marksmen at windows commanding egress from the barracks, and setting fire to the last, shoot the soldiers as they attempt to form.'[39]

Accordingly, Napier proposed keeping his forces concentrated, with some 900 men under Sir Hew Ross in Carlisle, Newcastle, Tynemouth and Sunderland; another 2,800 men under Colonel Thomas J. Wemyss in Manchester, Stockport, Bolton, Blackburn, Burnley, Todmorden, Rochdale, Wigan, Haydock, Liverpool and Chester; and a third force of 1,000 men based in Hull, York, Leeds, Sheffield, Derby, Nottingham and Halifax under his own command. He envisaged being able to support each subordinate force in strength and not dissipate his numbers, even using the railways where available, in response to requests from distraught magistrates. He insisted that if magistrates wanted detachments, they had to provide for the safety of the soldiers by providing 'a good barrack', as the magistrates at Bury had promised, or call upon local yeomanry.[40] After reviewing the temporary barracks in the north of England, Napier submitted a formal report, which recommended that fresh sites should be chosen near railways, roads and fresh water, and on the edges of towns, so that the troops could be deployed quickly yet preserved from sudden attack. By establishing large garrisons at places such as Thornhill in the West Riding of Yorkshire, he argued, soldiers could be kept 'out of mischief' while drill and discipline would prosper.[41]

If railway usage was still relatively limited in Napier's era, it was increasingly prominent during the Plug Plot disturbances of July–August 1842,[42] when regiments were reportedly whirled about 'at a rate of forty miles an hour',[43] and the renewed Chartist agitation of the mid- and late 1840s. A Railway Act, which received the royal assent only six days before the outbreak of the Plug Plot disturbances, contained a clause that compelled railway companies to convey soldiers at certain charges on presentation of an order signed by the proper authorities.[44] This proved to be a time when the authorities were able to move some 118,000 soldiers, and 12,000 dependants, by rail over two calendar years ending on 31 December 1843,[45] and commentators have hailed this transformation as providing 'a decisive edge' in the maintenance of public order.[46] The railways and their accompanying telegraphs, wrote the railway historian Jack Simmons, 'added immeasurably to the real power that could be exercised by the central government in London over the whole Kingdom'.[47]

Sir James Willoughby Gordon, then the quartermaster-general, was the source of this information on railway usage. Testifying before a parliamentary select committee on 1 March 1844, he famously remarked that the army could 'send a battalion of 1,000 men from London to Manchester in nine hours; that same battalion marching would take 17 days'. The men, he added, all long-service soldiers, some of whom were nearing the end of their twenty-one years of service, would arrive 'at the end of nine hours just as fresh, or nearly so, as when they started'. He asserted, too, that the railways enabled a relatively small army to act in a much more responsive way than it would otherwise have been able to do: 'you could not have done one-tenth part of the work that it was required' to do, 'and necessarily to do, in the year 1842'. Moving men with all their arms, ammunition and accoutrements, weighing about half a hundredweight (or 63.5 kg) per soldier, was, as he explained, much less burdensome by rail. Although they travelled in third-class coaches, the men had seats, and some railway companies provided covers for the coaches at no extra charge. It was just as feasible to move cavalry with their horses by rail, and overall the process was marginally cheaper than marching. Above all, as Gordon observed, rail movement facilitated the power of concentration at designated destinations, and maximized the time available for active duty (and did not waste it in travelling across country).[48]

The above use of the railways referred not merely to movements in connection with the maintenance of public order but also to all manner of regular troop movements, including units returning from overseas tours of duty and those moving to west-coast ports bound for Ireland or the colonies. The two modes of travel had advantages and disadvantages from a soldier's perspective: as a Gordon Highlander recalled, in reflecting upon his first march as a recruit in the 1830s, 'No railways at that time – the train's a grand marcher; the civilians were very kind to us sodgers – too kind sometimes in the matter o' drink.'[49] As the army relied upon public houses to provide sustenance on the line of march (so consuming the bulk of the daily expenditure, as Gordon admitted), the railways delivered these long-service soldiers in a reasonably fit and efficient condition.

So there were undeniable benefits to the use of railways, and the railway mania of the mid-1840s opened up much of the country for relatively rapid penetration by the military and/or the police. Yet railway movements were hardly a panacea. As the Horse Guards began moving larger bodies of soldiers by rail, often from the southern counties to the Midlands and the northern counties, the railway stations became the focal point for demonstrations, and the track and property of the railway companies became targets for attack. On 13 August 1842, when 700 Grenadier Guardsmen and artillery were due to leave from Charing Cross station to Birmingham, and thence to Manchester, where major disturbances had occurred, large crowds gathered (see Figure 1) and greeted them with hisses, groans and cries of 'Don't go

Figure 1 'Departure of Troops by the London and Birmingham Railway', *Illustrated London News*, vol. 1, no. 15 (20 August 1842), p. 232

and slaughter your starving fellow countrymen.'[50] Similarly when eighteen prisoners were taken in two omnibuses under escort by a file of 11th Hussars to Elland railway station near Leeds, a mob gathered at the station and along the route in protest. Although they failed to free the prisoners, they attacked the returning cavalry at Salter Hebble, injuring Mr William Briggs, the magistrate in attendance, and three of the ten troopers, one mortally. There were also recurrent fears of rail sabotage, and so railway workers and soldiers became involved in protecting railway property and guarding tunnels. Only the timely arrival of soldiers prevented crowds at Stockport and on the outskirts of Manchester from tearing up the track in August 1842.[51] In one week, *John Bull* described a bewildering array of rail movements from London: on Monday morning, a detachment of artillery was sent to the terminus at Euston Square 'and immediately forwarded by a special train to Manchester'; on Tuesday morning, the 'remaining portion' of the 34th (Cumberland) Regiment arrived from Portsmouth 'and in half an hour were *en route* to Manchester'; on Thursday morning, the 73rd Regiment '"took the rail"' from Euston Square 'for the disturbed districts in Yorkshire' and, in the afternoon, 32nd (Cornwall) Regiment 'proceeded by railway to Bradford', while a detachment of the 6th Foot (Royal Warwickshire) 'proceeded to the Southampton Railway, at Vauxhall, on their way to Gosport, for the purpose of relieving the troops ordered to the manufacturing districts'. The newspaper also reported troop movements from Chatham, Canterbury and Belfast.[52]

Yet the rapid despatch of soldiers to centres of public disorder did not always ensure that they overawed the crowds. Small detachments of the 72nd (later 1st Battalion, Seaforth) Highlanders, having travelled by rail to Manchester in August 1842, and thence on to Preston and Blackburn, found themselves under assault from stone-throwing mobs in both towns: in Preston on 13 August and Blackburn two days later. Ordered to open fire by local magistrates on both occasions, they inflicted casualties, including four fatalities in Preston, but dispersed the crowds in both encounters. During a more co-ordinated action on 16 August, the 72nd, under the command of Lieutenant-Colonel Charles, George James Arbuthnot, acted with the Lancashire Yeomanry and the 11th Hussars to rout a mob marching on Blackburn and seize seventy-five prisoners.[53] Wellington, who was once again commander-in-chief, reckoned that 'The affair at Preston ought to produce good consequences, but as far as I can judge even that affair was very badly managed . . . it is not advisable', he added, 'to show the troops, or even to call them from their quarters, unless it should be necessary.'[54] Commenting on the same events, Sir James Graham, the home secretary, assured Sir Robert Peel, the prime minister, that 'This state of things cannot last long; and I trust and believe, that we shall extinguish it.'[55] Ironically, Peel was more worried at this time about a rumour spread by a railway guard that the queen had been assassinated at Windsor – a rumour that 'spread like wildfire' within the capital and was quashed only by information from the next train. 'It is most improper', wrote the premier, that 'the Railway Guards' should bring 'such Reports.'[56]

Graham was soon proved right as the disturbances subsided amid divisions among the leadership, mounting arrests and demoralization among the participants, while the state managed to make and sustain a framework of stability and encourage a return to work. The disturbances collapsed with a relatively small amount of bloodshed, with possibly twelve deaths, including one soldier and three policemen.[57] Graham, nonetheless, resented the readiness with which local magistrates responded to incidences of localized disorder by requesting the despatch of the military (it even became customary for magistrates to keep a few empty railway coaches in sidings in case they should require military assistance). He urged these authorities to create and maintain efficient local constabularies and to call upon the military only in an emergency: as he informed James Herbert, the third earl of Powis, in July 1845, 'I cannot consent to Troops being at any time made a substitute for the Civil Power.'[58]

Both civil and military forces, though, would be stretched fully by the rural disturbances that erupted in the Welsh counties of Pembrokeshire, Carmarthenshire and Cardiganshire, known as the Rebecca riots. Triggered by the imposition of tolls on roads, but soon reflecting a broader range of agrarian, social, local and religious discontent, the 'Daughters of Rebecca' first made their appearance in Pembrokeshire on 13 May 1839, when groups of men dressed in women's clothes demolished a tollgate at Efail Wen, near Narbeth. Subsequent attacks followed that summer, and again in 1842, before they reached their crescendo in the summer of 1843, including attacks usually at night-time on high-rent landlords, bailiffs, unpopular individuals, isolated farms and workhouses, as well as the destruction of tollgates. The Rebeccaites also sent over a hundred letters and notices threatening further attacks. Railways were of scant assistance in policing this remote, rural region, where the South Wales Railway was not authorized until 1845 and the Chepstow and Swansea line not completed until 1850. Troops like the 4th Light Dragoons either rode into Carmarthen following a disturbance or, like the detachments of the 73rd Regiment of Foot (later the 2nd Battalion, the Black Watch), marched 45 km into Carmarthen in two five-hour stages. As soldiers were rotated throughout the region, their numbers fluctuated, but they were 1,800 strong under Colonel James Frederick Love (73rd Foot) by September 1843.[59] They supplemented the small numbers of special constables, pensioners and later Metropolitan policemen sent into the locality. Among the detachments of infantry were the 41st (Welch) Regiment, 75th Foot (later 1st Battalion, Gordon Highlanders), and the 76th (later 2nd Battalion, Duke of Wellington's Regiment), as well as Royal Marines, artillery and the Castlemartin Yeomanry Cavalry.

While Rebeccaites may have been deterred from attacking workhouses and small towns when these were under guard by the soldiery, they were not inhibited by the nightly cavalry patrols, and none of them was apprehended by the military. Graham was appalled by Love's practice of sending soldiers to crime sites 'on the following day. The Troops', he added, 'are thus constantly paraded before the people with apparent impotence and the authorities are brought into contempt.'[60] Appointing Major-General George Brown as the new commander may have altered the military tactics, but the operations petered out only when bribes elicited information for the authorities and informers turned in some of Rebecca's leaders. Graham also sought political solutions to the troubles with a combination of inquiries, proposed concessions and long-term reforms (ultimately involving the South Wales Turnpike Trust Amendment Bill, which received the royal assent as an Act of Parliament on 9 August 1844, and amendments to the Poor Law Act in 1844). He endeavoured, too, to bolster the regional forces of law and order; in July 1843, he secured the support of Carmarthenshire Quarter Sessions for the creation of a county constabulary (and somewhat belatedly, in January 1844, a similar resolution from the Cardiganshire Quarter Sessions).[61] So the state had a range of options that it could exercise in responding to rural disturbances, where the rapid deployment and/or movement of troops by rail did not exist.

Just as important were the precautions taken before industrial disruption or the larger disturbances occurred. Faced with the prospect of colliery strikes in Durham, initially at Thornley colliery in 1843, and then the 'great strike' of 1844, Sir James Graham was able to reassure Charles William Vane (formerly Stewart), the third marquess of Londonderry, that military forces would be retained in the region, and that any strike 'must be watched with vigilance and care'.[62] He was hugely relieved to learn that the 'great strike' had been broken by the 'judicious' arrest of strike leaders and delegates and the use of strike breakers – 'new Hands protected from violence'.[63] This was done without the expensive (and potentially provocative) recourse to Metropolitan policemen or the use of the military forces held nearby under the command of Lieutenant-General Sir Thomas Arbuthnot.

The latter, Graham averred, were be used only with 'great caution and forbearance'.[64] As in west Wales, he pressed the lord lieutenant to exhort his magistrates 'to make a permanent and efficient augmentation of their own Civil Force', which had been created in 1839, and by October 1845 he was delighted to learn that a 'beginning' had been made in increasing the size of the county's police force.[65]

Railways, in other words, may have been a 'force multiplier' in the process of suppressing industrial and political disturbances, but they only emphasized the need to plan ahead and to use the military, if at all, as an instrument of last resort. Vindicating these precepts were the events of 1848, following the stimulus provided by the revolutionary events in France and the localized rebellion in Ireland.[66] Facing another upsurge of Chartism, with a vast demonstration planned for Kennington Common, London, on 10 April 1848, and the presentation of another petition demanding the implementation of the People's Charter, the authorities needed little reminding that the overuse of the military could produce unwelcome outcomes. Another five people had been killed as recently as 6 March 1848, when the military were ordered to open fire on a mob during the bread riot in Glasgow. Accordingly, in advance of the London demonstration, vast numbers of special constables were raised, pensioners were enrolled, and precautions were taken to allow the meeting provided that it remained peaceful. These precautions included the guarding of public buildings, lighting London's gas lamps early to enhance security, stationing police at various strategic points, especially at the bridges across the Thames, and keeping the military out of sight. Ultimately, the Whig government assembled 7,122 military, including cavalry, for the defence of London, as well as 1,231 pensioners, over 4,000 Metropolitan and city police and possibly 85,000 special constables. The military, as Wellington advised, 'if not required it would be best not to shew them! and in the meantime their position will preserve in security all the essential points of the Town, the great communications between each of the military stations; and with Head Quarters and the Government and Parliament'.[67] Among the places chosen for protection were the goods station and depot at Nine Elms, thereby protecting the London and South Western Railway's line to the Solent ports.[68]

Railways, if not decisive in and of themselves, had proved their value as an adjunct in the maintenance of public order. They would remain critically important whenever disorder erupted in the more distant parts of the realm, whether in Exeter (1867), Skye (1883) or the miners' strike in the Rhondda Valley (1910).[69] They were used, too, in Ireland, where in response to the gun-battle fought by Irish nationalists and Irish police at Ballingarry (29 July 1848), the 75th Foot travelled by rail to Bagenalstown on 31 July, and then marched on to Kilkenny before advancing another 32 km to Ballingarry. For another six weeks they served in the field force under Major-General John MacDonald.[70] As the Irish rail network expanded, so did hopes that this would provide a crucial element in containing nationalist revolts. However, when faced with the Fenian revolt of 1867, the commander-in-chief of the British forces in Ireland, Sir Hugh Rose, the first Baron Strathnairn, realized that he had been too optimistic about the potential contribution of the railway. On account of the 'extraordinary powers' and 'rapidity' of the railway, he wrote, 'I lost sight of . . . this line of Rail, long as it is, which can transport nothing laterally, & of the vast extent of ground not under the influence of Rail, over which animals must in spite of water or Rail Transport, still transport all the wants of an Army in the field.'[71] So railways had raised possibilities for the British army, but the army had to be organized to exploit them fully.

Notes

1 W. T. Jackman, *The Development of Transportation in Modern England*, 2 vols. (Cambridge: Cambridge University Press, 1916), vol. 2, p. 658.

2 P. S. Bagwell, *The Transport Revolution from 1770* (London: Batsford, 1974); W. Schivelbusch, *The Railway Journey: The Industrialisation of Time and Space in the Nineteenth Century* (Leamington Spa: Berg, 1986).

3 P. J. Marshall (ed.), *The Cambridge Illustrated History of the British Empire* (Cambridge: Cambridge University Press, 1996), p. 117.

4 R. Kubicek, 'British Expansion, Empire and Technological Change' in A. Porter (ed.), *The Oxford History of the British Empire*, vol. 3: *The Nineteenth Century* (Oxford: Oxford University Press, 1999), p. 257.

5 Railway historians will rightly note that Middleton colliery line at Leeds was the first line sanctioned by Act of Parliament in 1758; the French inventor Bob-Joseph Cugnot built a prototype of a steam locomotive as early as 1769; and the Cornishman Richard Trevithick produced the first locomotive to run on a tramway in South Wales in 1804. M. Robbins, *George & Robert Stephenson* (London: Her Majesty's Stationery Office, 1981), pp. 9 and 15; Lt.-Col. E. W. C. Sandes, *The Military Engineer in India*, 2 vols. (Chatham: Institution of Royal Engineers, 1933), vol. 1, p. 104.

6 T. R. Gourvish, *Railways and the British Economy 1830–1914* (Manchester: Manchester University Press, 1980), p. 12; M. Freeman and D. Aldcroft, *The Atlas of British Railway History* (London: Croom Helm, 1985), p. 20; W. J. Ashworth, 'Industry and Transport' in C. Williams (ed.), *A Companion to Nineteenth-Century Britain* (Oxford: Blackwell, 2004), pp. 223–37, at p. 233.

7 'Chartism' is the umbrella name for numerous loosely co-ordinated local groups, often named 'working men's associations', articulating grievances in many cities from 1837 onwards. On policing, see J. M. Hart, 'The reform of the borough police, 1835–56', *English Historical Review*, vol. 70 (1955), pp. 411–27, at p. 416; F. C. Mather, *Public Order in the Age of the Chartists* (Manchester: Manchester University Press, 1959), pp. 113–15; and, on rise of the English police generally, C. Emsley, *The English Police: A Political and Social History*, 2nd edition (Harlow, Essex; Longmans, 1996).

8 R. Paley, '"An imperfect, inadequate and wretched system?" Policing London before Peel', *Criminal Justice History*, vol. 10 (1989), pp. 95–130.

9 H. Shore, 'Crime, Policing and Punishment' in Williams (ed.), *A Companion to Nineteenth-Century Britain*, pp. 381–95, at p. 388; and on Scotland, see K. Carson and H. Idzikowska, 'The Social Production of Scottish Policing, 1795–1900' in D. Hay and F. Snyder (eds.), *Policing and Prosecution in Britain, 1750–1850* (Oxford: Clarendon Press, 1989), pp. 267–97, and D. G. Barrie, 'A Typology of British Police: Locating the Scottish Municipal Police Model in its British Context, 1800–1835', *British Journal of Criminology*, vol. 50 (2010), pp. 259–77.

10 The People's Charter (1838) demanded political reform in respect of the vote for every man over the age of twenty-one years, the abolition of any property qualification for Members of Parliament, annual parliaments, use of the ballot, payment of Members of Parliament and constituencies of equal size. Many of these reforms later occurred but not because of the activities of the Chartists directly.

11 S. H. Palmer, *Police and Protest in England and Ireland 1780–1850* (Cambridge: Cambridge University Press, 1988), p. 415.

12 F. O. Darvall, *Popular Disturbances and Public Order in Regency England* (London: Oxford University Press, 1969), p. 260; see also I. F. W. Beckett, 'The Militia and the King's Enemies, 1793–1815' in A. J. Guy (ed.), *The Road to Waterloo: The British Army and the Struggle against Revolutionary and Napoleonic France, 1793–1815* (Stroud, Gloucestershire: Alan Sutton for the National Army Museum, 1990), pp. 32–9, at p. 35, and Maj. O. Teichman, 'The Yeomanry as an Aid to Civil Power, 1795–1867', *Journal of the Society for Army Historical Research*, vol. 19 (1940), pp. 75–91 and 127–43, at pp. 78–9.

13 The first organized police forces in Ireland came about through the Peace Preservation Act (1814) and then the Irish Constabulary Act (1822), which established a Constabulary Police (later re-designated as the Royal Irish Constabulary) across all the provinces of Ireland. By 1841, this centrally controlled, highly disciplined and armed constabulary numbered over 8,600 men, while Dublin acquired its own unarmed constabulary in 1836.

14 N. Higgins-McHugh, '4: The 1830s Tithe Riots' in W. Sheehan and M. Cronin (eds.), *Riotous Assemblies: Rebels, Riots and Revolts in Ireland* (Cork: Mercier Press, 2011), pp. 80–95; see also V. Crossman, 'The Army and Law and Order in the Nineteenth Century' in T. Bartlett and K. Jeffery (eds.), *A Military History of Ireland* (Cambridge: Cambridge University Press, 1996), pp. 358–78, at pp. 359 and 372–3.

15 They still acted in aid of the civil power, with the 2nd Life Guards sent to Brighton during the Sussex riots of 1830 and in action again during the Chartist disturbances of 1838–40: B. White-Spurner, *Horse Guards* (London: Macmillan, 2006), p. 355. In August 1842, 700 Grenadier guardsmen were also sent to Manchester: F. C. Mather, 'The Railways, the Electric Telegraph and Public Order during the Chartist Period, 1837–48', *History*, vol. 38 (1953), pp. 40–53, at p. 43.

16 E. M. Spiers, *The Army and Society, 1815–1914* (London: Longman, 1980), p. 74; C. T. Atkinson, *History of the Royal Dragoons 1661–1934* (Glasgow: R. Maclehose & Co., 1934), pp. 314–15.

17 I. F. W. Beckett, 'Introduction' and 'Britain' in I. F. W. Beckett (ed.), *Citizen Soldiers and the British Empire, 1837–1902* (London: Pickering & Chatto, 2012), pp. 1–21 and 23–40, at pp. 6 and 23.

18 Of the nineteen Scottish yeomanry regiments, only two survived (the Ayrshire and Lanarkshire, though the East Lothian Yeomanry reappeared in 1846), and after 1838, the sole remaining Welsh corps was the Pembrokeshire (Castlemartin) Yeomanry: Beckett, 'Britain' in Beckett (ed.), *Citizen Soldiers*, pp. 28–9.

19 R. Walmsley, *Peterloo: The Case Reopened* (Manchester: Manchester University Press, 1969); R. Reid, *The Peterloo Massacre* (London: Heinemann, 1989); R. Poole, 'By the Law or the Sword: Peterloo Revisited', *History*, vol. 91 (2006), pp. 254–76.

20 Parliamentary Debates (Parl. Deb.), Third Series, vol. 42, col. 651 (27 April 1838).

21 J. Saville, *1848: The British State and the Chartist movement* (Cambridge: Cambridge University Press, 1987), pp. 23–5.

22 G. A. Williams, *The Merthyr Rising* (London: Croom Helm, 1978), ch. 5; D. J. V. Jones, 'The Merthyr riots of 1831', *Welsh History Review*, vol. 3, no. 2 (1966),

pp. 173–205 and 'Law Enforcement and Popular Disturbances in Wales, 1793–1835', *Journal of Modern History*, vol. 42 (1970), pp. 496–523.

23 Lord Ellenborough, diary, 31 October 1831, in A. Aspinall (ed.), *Three Early Nineteenth Century Diaries* (London: Williams and Norgate, 1952), p. 153; see also Teichman, 'Yeomanry as an aid to Civil Power', pp. 130–4, and, on casualties, J. Stevenson, *Popular Disturbances in England 1700–1832*, 2nd revised edition (London: Longmans, 1992), p. 292.

24 E. Longford, *Wellington: Pillar of State* (London: Weidenfeld & Nicolson, 1972), p. 277; see also L. T. C. Rolt, *George and Robert Stephenson: The Railway Revolution* (London: Longmans, 1960), pp. 193–202.

25 This railway also set the precedent of agreeing with the government upon cheap fares for the troops; other railways followed suit in the early 1840s. J. Westwood, *Railways at War* (London: Osprey, 1980), p. 6; R. P. Dunn-Pattison, *The History of the 91st Argyllshire Highlanders* (Edinburgh: Blackwood and Sons, 1910), pp. 98–9.

26 Palmer, *Police and Protest*, p. 434; on the military districts, see Saville, *1848*, p. 24.

27 Mather, *Public Order*, pp. 105–7; Palmer, *Police and Protest*, p. 447.

28 Among the ex-military men who held important management railway posts were Captains J. M. Laws and C. H. Binstead (Manchester and Leeds), Captain J. E. Cleather (Grand Junction, Manchester and Birmingham), Captain J. W. Coddington (Caledonian), Captain W. O'Brien (Great North of England) and Captain C. R. Moorsom and Lt. H. P. Bruyères (London and Birmingham), but the most famous was Captain Mark Huish, who served as secretary of the Glasgow, Paisley and Greenock Railway (1837–41) and then as secretary and general manager of the Grand Junction Railway (1841–46) before becoming general manager of London and North Western Railway (1846–58). T. R. Gourvish, *Mark Huish and the London and North Western Railway: A Study of Management* (Leicester: Leicester University Press, 1972), pp. 27–8, and M. R. Bonavia, *The Organisation of British Railways* (Shepperton: Ian Allan, 1971), pp. 13–14. On the militarized culture that evolved within the Victorian railways, see J. Richards and J. M. MacKenzie, *The Railway Station: A Social History* (Oxford: Oxford University Press, 1986), pp. 98–9.

29 Mather, 'Railways, the Electric Telegraph and Public Order', pp. 47–8.

30 Lt.-Gen. Sir W. Napier, *The Life and Opinions of General Sir Charles James Napier, G.C.B.*, 4 vols. (London: John Murray, 1857), vol. 2, p. 44. The other units were the 96th Foot, 2nd Dragoon Guards and 8th Hussars. Mather, *Public Order*, p. 161 n.1.

31 Napier, *Life*, vol. 2, p. 39; N. R. Pye, 'The Home Office and the Suppression of Chartism in the West Riding, c. 1838–1848', unpublished thesis submitted towards a PhD (University of Huddersfield, 2011), p. 195; Mather, *Public Order*, pp. 162–3.

32 The National Archives (TNA), HO 50/16, 'Distribution of the Army in Great Britain', 1 February 1840; see also Mather, *Public Order*, p. 163.

33 D. J. V. Jones, *The Last Rising: The Newport Insurrection of 1839* (Oxford: Clarendon Press, 1985), pp. 146–55, 160–2, 166–7.

34 H. Strachan, *The Politics of the British Army* (Oxford: Clarendon Press, 1997), p. 83; Mather, *Public Order*, p. 154; Napier, *Life*, vol. 2, pp. 4, 9, 12, 23, 39, 51.

35 Napier, *Life*, vol. 2, p. 63.

36 Ibid. pp. 15, 24–5, 40.

37 Ibid. pp. 4, 7–8, 15–16, 47; see also B[ritish] L[ibrary], Add. MSS 49129, Napier MSS, Napier to Williams, 9 April 1839.

38 Napier, *Life*, vol. 2, pp. 32, 49, 54, 60.

39 Ibid. p. 47. There was never any evidence of Lancastrian Chartists planning nocturnal attacks and selective assassination; these fears reflected Napier's memories of Captain Swayne and his Cork militiamen being burnt alive in their barracks in 1798, BL, Add. MSS 49129, Napier MSS, Napier to Col. Wemyss, 10 April 1839, and Napier to S. Phillips, 13 April 1839; Palmer, *Police and Protest*, p. 435.

40 Napier, *Life*, vol. 2, pp. 8, 17, 19–20, 30–1, 35–6; TNA, HO 50/16, Horse Guards to S. Phillips, 27 July 1840, enclosing extract from Napier on barrack conditions.

41 BL, Add. MSS 54515, Napier MSS, Sir C. J. Napier, 'With regard to a central barrack for 800 men at Brighouse or Thornhill Briggs', 1840; see also H. Strachan, *Wellington's*

Legacy: The Reform of the British Army, 1830–54 (Manchester: Manchester University Press, 1984), p. 61.

42 Sometimes described as a general strike, the Plug Plot disturbances derived their original name from the fact that mills were prevented from working by the removal or 'drawing' of bolts or 'plugs' from the boilers, so preventing steam from being raised. See also F. C. Mather, 'The General Strike of 1842: A Study in Leadership, Organisation and the Threat of Revolution during the Plug Plot Disturbances' in R. Quinault and J. Stevenson (eds.), *Popular Protest and Public Order: Six Studies in British History* (London: George Allen & Unwin, 1974), pp. 115–35.

43 T. Raikes, quoted in Longford, *Wellington*, p. 432.

44 Parliamentary Papers (PP), General Sir James Willoughby Gordon, evidence (1 March 1844), included in *Fifth Report from the Select Committee on Railways: 1844* (1844), XI, p. 144.

45 Ibid.

46 Palmer, *Police and Protest*, p. 409; see also Saville, *1848*, pp. 24–5.

47 J. Simmons, *The Victorian Railway* (London: Thames and Hudson, 1991), p. 364.

48 PP, Gordon, evidence, pp. 144–8.

49 Lt.-Col. G. Greenhill Gardyne, *The Life of a Regiment: The History of the Gordon Highlanders from 1816 to 1898*, vol. 2 (London: The Medici Society, 1903), p. 37.

50 'The Disturbed State of the Country', *Manchester Guardian*, 17 August 1842, p. 1.

51 'Attack upon the Military at Salterhebble', *Leeds Mercury*, 20 August 1842; see also Palmer, *Police and Protest*, p. 459; 'The Manchester Riots', *John Bull*, 13 August 1842, p. 395, and 'Disturbances in the Manufacturing Districts', *John Bull*, 20 August 1842, p. 399.

52 'Disturbances in the Manufacturing Districts', *John Bull*, 20 August 1842, p. 399.

53 Lt.-Col. A. Fairrie, *'Cuidich 'N Righ': A History of the Queen's Own Highlanders (Seaforth and Camerons)* (Inverness: Regimental Headquarters, 1983), p. 8; BL, Add. MSS 40047, fo. 70, Peel MSS, Sir James Graham to Sir Robert Peel, 16 August 1842; Blackburn Library, Charles Tiplady, diary, microfilm copy, August 1842.

54 The Duke of Wellington to Sir J. Graham, 15 August 1842' in C. S. Parker (ed.), *Life and Letters of Sir James Graham Second Baronet of Netherby, P.C., G.C.B. 1792–1861* (London: John Murray, 1907), pp. 325–6.

55 BL, Add. MSS 40047, fo. 70, Peel MSS, Graham to Peel, 16 August 1842.

56 BL, Add. MSS 40047, fo. 72, Peel MSS, Peel to Graham, 16 August 1842.

57 Palmer, *Police and Protest*, pp. 459, 461.

58 TNA, HO 41/18, Sir J. Graham to Earl Powis, 30 July 1845. This became a recurrent theme of messages from the Home Office: see TNA, HO 41/18, Home Office to the mayor of Penzance, 28 May 1847, and A. Babington, *Military Intervention in Britain from the Gordon Riots to the Gibraltar Incident* (London: Routledge, 1991), p. 111.

59 BL, Add. MSS 40449, fo. 37, Peel MSS, Graham to Peel, 17 September 1843; see also D. Williams, *The Rebecca Riots: A Study in Agrarian Discontent* (Cardiff: University of Wales Press, 1955), pp. 213, 266.

60 BL, Add. MSS 40449, fo. 38, Peel MSS, Graham to Peel, 17 September 1843.

61 Williams, *Rebecca Riots*, pp. 224, 226, 266; D. J. V. Jones, *Rebecca's Children: A Study of Rural Society, Crime, and Protest* (Oxford: Clarendon Press, 1989), pp. 223, 225, 238, 346–9, 352–3. On Graham's often fraught relations with county magistrates and his attempts to blackmail them into forming county police forces by withholding the deployment of military forces, see A. P. Donajgrodzki, 'Sir James Graham at the Home Office', *Historical Journal*, vol. 20, no. 1 (1977), pp. 97–120, at pp. 105, 110–12.

62 Durham County Record Office (DRO), D/Lo/C 80, Londonderry MSS, Sir James Graham to Lord Londonderry, 6 June 1843 and 17 April 1844. Londonderry had multiple interests in these security measures. A distinguished former cavalry commander, popularly known as 'Fighting Charlie', he was also a politician and diplomat (a half-brother of Lord Castlereagh), and had married into the extremely wealthy Vane family, becoming a prominent mine-owner as well as the lord lieutenant of County Durham.

63 Ibid. Graham to Lord Londonderry, 29 May 1844.
64 Ibid. Graham to Lord Londonderry, 14 June 1844 and 23 July 1844.
65 Ibid. Graham to Lord Londonderry, 14 June 1844 and 19 October 1845.
66 The Irish constabulary prevailed in this conflict, winning the gun-battle known as the 'battle of Ballingarry' (29 July 1848) and later arresting several leaders of the Young Irelander movement, while others escaped to France.
67 TNA, WO 30/81, 'Memorandum by the Duke of Wellington', 5 April 1848: on the numbers raised by the authorities, see Saville, *1848*, p. 109, and TNA, WO 30/111, 'Chartist Riots'.
68 M. Chase, *Chartism: A New History* (Manchester: Manchester University Press, 2007), pp. 330–1.
69 Simmons, *Victorian Railway*, pp. 365–6; see also E. A. Cameron, 'Internal Policing and Public Order, c. 1797–1900' in E. M. Spiers, J. A. Crang and M. J. Strickland (eds.), *A Military History of Scotland* (Edinburgh: Edinburgh University Press, 2012), pp. 436–57, at p. 450.
70 Greenhill Gardyne, *Life of a Regiment*, vol. 2, pp. 212–13.
71 BL, Add. MSS 42823, fo. 166, Rose MSS, Baron Strathnairn to Grant, 24 March 1867.

2

Railways and home defence

If the early Victorian army had struggled initially to discharge its duties in aid of the civil power, and welcomed railways as an ancillary means of responding to domestic emergencies, it encountered similar challenges in fulfilling its role in home defence. As in aiding the civil power, the army's role in home defence was strictly a subordinate one. Prime responsibility for this task rested with the Royal Navy, whose dominance of the high seas, including the English Channel, had been uncontested since the battle of Trafalgar (21 October 1805). In an age of sail and wooden-hulled warships the Senior Service not only remained the world's dominant naval power but also enjoyed political and public support, especially the endorsement of a large, literate and nationalistic section of middle-class opinion. It also had access to the financial, scientific, industrial and organizational resources to meet any external challenge. Nevertheless, invasion scares occurred, particularly during the middle of the nineteenth century, followed by security controversies aroused by Channel Tunnel schemes. All these episodes raised concerns about the capacity of Britain's military forces to mobilize and concentrate sufficient forces to repel large-scale invasions.

Invasion scares

The three mid-century invasion scares (1846–48, 1851–52 and 1857–58) reflected anxieties about how France could exploit technological developments in steam navigation, shell artillery and ironclad warships. They generally involved scenarios whereby the Royal Navy was lured away from the Channel, so enabling large invasion forces to be landed on the southern shores of England. In July 1845, Lord Palmerston had famously declared that the development of steam power had reduced the Channel to 'nothing more than a river passable by a steam bridge'.[1] Barely a year previously François d'Orléans, the prince de Joinville and third son of the French king, Louis Philippe, had conducted naval operations off the coast of Morocco, bombarding Tangier (6 August

1844) and occupying Mogador (16 August 1844). Avowedly anti-British, the prince campaigned for the creation of a steam navy for France, and although the French naval law of 1846 never represented an immediate threat to Britain, it contributed to the first invasion 'panic'.[2] That panic did not subside until Louis Philippe and his family fled from the revolution of February 1848 and settled in England. The second 'panic' erupted in 1851 after Louis Napoléon, who had been elected president of the Second Republic in 1848, followed in his uncle's footsteps and established the Second Empire. Crowned as Napoleon III, he became emperor of the French on 2 December 1852 (the forty-eighth anniversary of his uncle's coronation as Napoleon I). This invasion scare soon subsided, too, as Britain and France found themselves allied with the Ottoman empire, and later with the kingdom of Piedmont-Sardinia, during the Crimean War (1854–56). Despite an Anglo-French naval arms race in steam-powered battleships during the 1850s, the third invasion scare followed the decision of France to begin work on the first 'ironclad' warship, *La Gloire*, in 1857, which was eventually trumped by the launching of HMS *Warrior*, the first armour-plated, iron-hulled warship, in December 1860.

To confront any anticipated invasion, the British army languished under the same handicaps that had jeopardized its capacity to aid the civil power. The army at home was simply far too small, with its numbers depleted by garrison commitments in Ireland and deployments in areas blighted by industrial or agrarian disturbances. It lacked the support of auxiliary forces other than the yeomanry, and encountered a profound unwillingness on the part of successive governments to increase the army estimates significantly. Meeting the Chartist threat had required some increases but these were relatively marginal, and the annual expenditure on the ordnance increased only from £87,210 in 1843 to £294,842 in 1847. Moreover, it was the Royal Navy and not the army that tended to benefit from the recurrence of crises in Anglo-French relations.[3]

Nevertheless, on 10 September 1845, when the duke of Wellington apprised Sir Robert Peel, then prime minister, about the state of the nation's defences, rail connections were of pivotal importance for him. He claimed that the nation's security had 'greatly deteriorated', and that key points on the south coast, namely Plymouth, Portsmouth and Dover Castle, had to be garrisoned. In responding to any threat, half of the disposable force had to be assembled in London at 'the central terminus of each of the Rail Roads in Great Britain', while the other half had to be assembled in the neighbourhood of Dublin in the first instance 'as the central terminus of the Rail Roads in Ireland'. Rail connections were already central to his assumptions: 'Having prepared & organized the means of sending horses as well as men by Rail Roads I would send off to any part menaced a sufficient force to defeat any that might be landed by the enemy.'[4]

The duke was only one of several senior military officers concerned about the state of Britain's land defences. He informed Peel of the anxieties of Sir George Murray, who had served the duke as his quartermaster-general for the first four years of the Peninsular War (1808–14) and was now, at seventy-three years of age, the master-general of the ordnance. Murray feared that London, then in an 'open and defenceless condition', would remain the most likely object of an invasion; accordingly, it would remain 'the temptation to such undertaking', and that temptation would be 'very strong'.[5] Murray had already received 'the first authoritative warning'[6] from another Peninsular veteran, Major-General Sir John Fox Burgoyne, then a mere sixty-three years of age and newly appointed as the inspector-general of fortifications. Burgoyne had sent his paper to Murray, dated 11 September 1845, in which he intimated that railroads could move up to 400 men in any one train, and that ten trains could be prepared quickly and 'sent off in quick succession', provided that suitable preparations were made. 'Railways', argued Burgoyne, 'will be of great service in affording rapid means for concentrating troops at any threatened point, or near one that is attacked', although there might be a 'great deficiency of appropriate carriages'. He made the case for investigating how carriages could be 'most readily fitted out' for carrying troops, artillery and ammunition, for making an inventory of the 'wheels, axles & springs' and for calculating the number of trains required to carry this force at 'a velocity of 24 miles [36 km] an hour'.[7]

Burgoyne's views on railways represented a sharp departure from those of his predecessor, General Sir Frederick Mulcaster (inspector-general of fortifications from July 1834 to July 1845), the Board of Ordnance and some senior engineers including Colonel M. Dixon, the Commanding Royal Engineer (CRE), Jersey. During the 'railway mania' all these officials had been bombarded with proposals for railway developments in their areas. The solicitor of the Ordnance Board had objected repeatedly to various railway proposals in Kent and Sussex,[8] and even in respect of the proposed London, Chatham and Dover railway in 1836, where the CRE, Dover, could find no objections from 'a military point of view', Mulcaster had warned against any buildings, embankments or other works 'that may obstruct the fire or injure the fortifications of Dover'.[9] In Jersey, too, where the island's proximity to France exacerbated fears of an invasion,[10] local officers of the Royal Jersey Militia, Colonels Thomas le Bruton and John Le Couteur, submitted a petition, arguing the strategic case for developing an island railway. They asserted that a railway would afford a 'quick means' of moving soldiers to oppose a landing, and that the construction of a harbour in the vicinity of Gorey (the closest headland to France) would enable a steamer to carry intelligence of an impending French assault to Southampton in eight hours. Colonel Dixon, though, feared that the railway could be turned into a 'weapon of offence by an enterprising enemy'; and, unlike the local militia colonels or Burgoyne, he did not believe in a disposable force moving to confront an invader. Instead he envisaged a fixed-site defence of the island, with all available forces divided and retained in forts to watch for an enemy landing.[11]

Burgoyne had developed an early interest in railways; in 1836, he had served as a member of a royal commission to consider and recommend a general system of railways for Ireland, and he was generally credited with writing much of the final report.[12] His invasion anxieties, dovetailing as they did with those of the duke of Wellington, underscored the need for a disposable force or mobile reserve that could be transported rapidly to confront an adversary on the southern coast. In 1846 Burgoyne gave evidence before a royal commission on railway gauges,[13] and amplified his calculations of the ability to move soldiers and their equipment by railway. 'I look upon the whole safety of the kingdom to depend upon railways,' he affirmed. Given the capacity of France to utilize steam vessels to land large bodies of troops on the southern coast, Burgoyne maintained that 'nothing but the power of concentration which the railways would give could enable you to resist successfully'. As soldiers no longer had to be kept 'within reach of every port', they could be based (as many were in the mid-1840s) '200 miles [322 km] in the north'. They could be brought south by rail 'within 24 or 48 hours', so giving 'the great advantage of which is well known in military tactics, the power of concentrating upon any given point in a short time'. This could prove crucial, argued Burgoyne, if Britain deployed defensive forces before the enemy could land 20,000 men and all their equipment (which the British knew from past experience could take several days or even weeks to accomplish). In this race to concentrate troops 'with the greatest rapidity, and in the greatest numbers', he did not doubt that Britain had sufficient locomotives to transport '100,000 men in a short period' (some ten times above the number of men currently available), but the number of carriages would still be 'very great for 3,600 men; 150 carriages would be wanted', of which 120 had to be adapted to carry soldiers and their equipment. Burgoyne reckoned that it would not be cost-effective to move large numbers of cavalry by train – '90 horses would require 30 carriages; 900 would require 300 carriages' – but a brigade of artillery, requiring 120 or 130 horses, could be moved by train.[14] While Burgoyne was less concerned than the quartermaster-general, Sir J. Willoughby Gordon (and the commissioners), about the inconvenience of having to change from one line to another on account of differing gauges, he supported another defensive measure, namely the creation of a coastal railway line from Portsmouth to Plymouth.[15]

However, Burgoyne's real fear lay not with the ability to move a disposable force but with the size of that force, which in his opinion barely numbered 5,000 to 10,000 regular troops in 1846. Unconvinced by the naval alternative of fitting out the reduced hulls of men-of-war as floating steam batteries, Burgoyne was equally appalled by the lack of adequate stores and equipment for any disposable force and the lack of fortresses protecting key harbours along the southern coast. Whereas France, he reckoned, could 'by partial movements, scarcely to be observed, collect from 100,000 to 150,000 troops on the shores of the Channel, within a few hours' sail of the British coast', and possess sufficient vessels and steamers to deploy the force, 'England would have neither the fortifications nor troops, nor means of equipment for a force equal to cope with even one fourth of that number.' He duly recommended that the regular army should be increased significantly. This would enable a force of 30,000 men, held in constant readiness, to take the field at once (and be doubled in an emergency to 60,000), with fortifications constructed 'at every port in proportion to its importance'.[16]

Burgoyne's paper, once printed and circulated, attracted the attention of Lord Palmerston, once again the foreign secretary in the Whig government of Lord John Russell. Palmerston prepared a paper for the cabinet, alluding to the 'defenceless state of the country', and added to the urgency of the case by observing that France was constructing railways 'by the friendly aid of British capital, which will soon give the French great additional facilities for the transport of men and stores'. He endorsed all of Burgoyne's requirements, recommending the re-creation of a militia, not raised by ballot in the first instance but capable of serving as 'an army of reserve'.[17] Having already championed the cause of increasing the regular army and raising the militia in the previous April, Palmerston would do so again, with the support of the duke and his paper based upon Burgoyne's calculations. Yet none of these interventions persuaded the cabinet.[18]

When Burgoyne's memorandum was sent to Wellington, he replied with a lengthy letter, dated 9 January 1847, endorsing Burgoyne's concerns and complaining that he had tried 'in vain . . . to awaken the attention of different administrations to this state of things'.[19] Although this letter after extensive private circulation found its way, via the indiscreet hands of Sir John's wife and daughters,[20] into *The Times* of

1 December 1847, and later extracts appeared in the *Morning Chronicle* of 6 January 1848, the ensuing public controversy still failed to produce any significant change of policy. The modest increase in the army estimates reflected continuing concern over the Chartists and not the invasion panic. Lord John Russell, the premier, was amenable to the arguments of Palmerston, Wellington and other militia advocates, but his proposal, as presented to parliament in January 1848, to raise a 200,000-strong militia required an increase in income tax from 7*d* to 1*s* in the pound. This was bound to provoke opposition, but as news of the revolution in France and the fall of Louis Philippe occurred in the following month, the public alarm duly subsided.[21]

The defenceless state of Britain, nonetheless, had aroused enduring anxieties, not the least of which was the need to concentrate army units by railway and train them in 'camps of exercise', so that they could learn their duties in combined bodies. Having raised this proposal with Wellington in 1847, Prince Albert[22] received regular briefings on rail-related issues from the Horse Guards. These included a memorandum on the application of railroads to military purposes, including detailed computations on the numbers of men, horses and ordnance that could be transported in railway carriages,[23] and copies of correspondence between the duke and several railway companies on their capacity to convey soldiers and their equipments.[24] Lord Hardinge, who succeeded Wellington as commander-in-chief of the army on 28 September 1852, shared his fears of invasion and, with the prince's support, revived the idea of 'camps of exercise'. In a memorandum of 10 November 1852, Albert envisaged the defence of key fortifications with regular, pensioner and militia forces, supported by mobile reserves concentrated at Reading on the Great Western Railway, Farnborough on the South Western, and Reigate on the South Eastern and Brighton lines.[25] This concerted pressure at least persuaded Sidney Herbert, as secretary at war, to include a vote in the army estimates to cover the expenses of a summer camp for 7,000 men.[26]

Another legacy of the first panic was the generation of all manner of schemes to protect the country. In 1848 William Malins, a Peninsular veteran and railway enthusiast, submitted *A Plan for Additional National Defences* to the duke of Wellington. While he advocated preparation 'at all points', his first priority was the coastline from Portsmouth to Dover, where improvements in rail and broad-gauge carriages as used on the Great Western could provide 'rapid concentration of the most powerful batteries'. This could provide 'an economical, complete, and perfect means of defence for our citadel island', he affirmed.[27] Lord Hardinge received a similar plan from James Anderson of Edinburgh. Anderson supported the creation of another

161 km of railway to assist in the defence of the southern coast, but his main suggestion was the creation of an ordnance carriage, made of malleable iron, that would serve as a gun carriage. Able to revolve like a railway turntable, this carriage could absorb the recoil at discharge, while the gun fired from an elevated or depressed trajectory by means of a screw attached to the axle of the carriage. Anderson claimed that this concept had met with the approval of colonels of the Royal Engineers and Royal Artillery in Edinburgh, and he envisaged the construction of 200 of these carriages, with the formation of four 'parks of flying artillery, each consisting of fifty of these ordnance carriages'. As the cost of labour and material had fallen sharply since 1849, Anderson reckoned that the cost of 100 miles (161 km) of rail and 200 of his ordnance carriages would be a mere £475,000.[28]

Such ingenuity was not put to the test. The second 'panic', following Louis Napoléon's coup of December 1851, prompted fierce parliamentary debate but primarily around the manner of increasing the size of the army. While the Whig government agreed to increase the regular army by 5,000 men, its militia proposals encountered strong opposition led by Palmerston, who had recently lost office as foreign secretary and who favoured a different militia scheme. In his 'tit for tat' with Lord John Russell, Palmerston brought down the government and then supported its Tory successor, under Lord Derby, when it sought to raise 80,000 militiamen in England and Wales by voluntary enlistment and the offer of bounties. This militia force, which could number 120,000 in case of a great emergency, would be paid and receive twenty-one days' training a year. During its passage through parliament, the Militia Bill attracted over 800 petitions against its provisions, and required some 200 hours of debate and thirty-two divisions before passing its second reading on 26 April 1852 and receiving royal assent on 30 June 1852. Similar legislation was later extended to Scotland (11 August 1854), where another 10,000 militiamen were to be raised.[29]

Burgoyne regarded the militia as a key component possibly ranging from one or two thirds of any army of the reserve, and so enabling 100,000 men to be deployed in the field. While the chances of success, he claimed, would vary with the proportion of regulars employed, he exuded confidence in view of 'the advantages that the mass of railways afford for rapid concentration at any point'.[30] Burgoyne was also a firm believer in the merits of a coastal railway line from Portsmouth through Dorchester to Exeter. Both the Dorchester and Exeter, and the Great Western, companies proffered proposals to the master-general of the ordnance, but the economic attractions were slight, and the lukewarm enthusiasm for the militia as the 'panic' began to ebb hardly sustained the option. The line was never built, and the issue was closed in 1856 when parliament approved another railroad further inland, which was more attractive commercially but had no military value at all.[31]

Finally, Burgoyne was one of the first senior officers to look beyond the militia and advocate the reform of the volunteers. In a memorandum sent to Lord Hardinge in May 1852 and later published as 'Militia and Volunteers' in the *United Service Magazine* (March 1853), he maintained that a reconstitution of the volunteers would afford a considerable means of military expansion in England and Wales, especially if the volunteers were armed with the new Minié rifle, commanded where possible by old regular officers, and organized as companies attached to regular regiments.[32] The idea gained traction after the Crimean War as Britain found its military resources stretched during the response to the mutiny in India,[33] and developments across the Channel, including the building of *La Gloire*, seemed increasingly ominous. Anxieties mounted over the French development of the port of Cherbourg, the allegations in France that the bomb thrown by Felice Orsini at Napoleon III was made in Birmingham, and fears that France would seek an alliance with Russia. British commentators were also suspicious of Napoleon's interference in Italy, which led to the annexation of Nice and Savoy in 1860. The unfolding panic spread across official circles and among sections of the press and public; it focused initially upon naval inadequacies, but soon encompassed calls for additional fortifications and the creation of Volunteer Rifle Corps.[34] *The Times* was a strong champion of the volunteers, criticizing hesitation in the Horse Guards, and ready to publish Alfred Tennyson's famous poem 'Riflemen Form' on 9 May 1859.[35]

Volunteers and railways

In authorizing the formation of Volunteer Rifle Corps three days later, the government was merely defusing external pressure in a relatively inexpensive manner. It stipulated that the force was liable to be called out in the event of an actual invasion or appearance of an enemy on the coast or in the case of rebellion. It also insisted that volunteers could be described as 'effective' only if they did twenty-four days' drill a year (as derived from the legislation of 1804 but under the new volunteer legislation of 1863, the efficiency test was reduced to merely nine drills and attendance at the annual inspection). Although the volunteers were not organized initially as a body that could be assembled rapidly for home defence, 700 corps were formed by 1861, embodying 161,000 volunteers,[36] and they had connections with the railways from the outset. Unlike many of the early corps raised by professional bodies, the 20th Middlesex Rifle Volunteers were known as the 'Railway Rifles' because they were raised from the workers of the London and North Western Railway Company with headquarters at Euston Square.[37] Among the other railway companies subsequently formed were the Crewe Rifles and the Swindon Volunteer Rifles, both subsidized with donations and the provision of accommodation from their respective railway companies, as well as the Great Eastern Essex Rifles and the Montgomeryshire (Railway) Rifles.[38] These artisan volunteers were highly regarded as respectable citizens who were suitably deferential towards authority. They knew that dismissal from the volunteers could lead to dismissal from their employers such as the London and North Western Railway Company in Buckinghamshire.[39]

Finally, the rifle volunteers received favourable terms from the railway companies, which competed for the custom of carrying them to their Easter reviews at least until the passing of the Bank Holidays Act (1871), when rail companies had so much civilian traffic that they could refuse to convey the volunteers. Yet special arrangements were still made, not least in Scotland, where the marches of volunteers to railway stations, their assembly at those stations and their subsequent return often proved memorable occasions, reviving displays of popular enthusiasm for the volunteer movement. Brigadier J. H. A. Macdonald fondly remembered the immense crowd that gathered at Waverley station in September 1872 to send 125 men of the Queen's Edinburgh Brigade off to manoeuvres on Salisbury Plain.[40] Great popular interest recurred when Queen Victoria reviewed the Scottish volunteers in Edinburgh in August 1881. The 1st Lanark Rifle Volunteers benefited from 'admirably conceived' and well-executed railway arrangements as they travelled through to Edinburgh, attracting massive crowds in Glasgow and Edinburgh, and later received extensive commendations from the local and military press.[41]

This relationship between the railway companies and the volunteers continued to evolve. Although the Crewe Rifles, which numbered 401 men in 1878, was disbanded two years later, it was replaced in 1887 by the formation of the Railway Engineer Volunteer Staff Corps. Conceived by Francis William Webb, the chief mechanical engineer of the London and North Western Railway Company, his subordinate officials and prominent local Conservatives, this corps soon consisted of six companies with twenty-four officers and 610 men. The volunteers were all employees of the railway company, comprising engine drivers, firemen, cleaners, boilermakers, riveters, fitters, smiths, platelayers, shunters and pointsmen. Later the corps was known as the 2nd Cheshire Engineer (Railway) Volunteers; 245 of its members enlisted as a matter of form in the regular army for one day, so that they could pass into the Royal Engineers Reserve, where they remained for six years, liable for active service. This unique force later saw active service in the South African War (1899–1902).[42]

Another product of the volunteer movement was the formation of the Engineer and Railway Volunteer Staff Corps in 1865. It consisted of officers only drawn from the ranks of civil engineers and contractors, officers of railway and dock companies and, under special circumstances, Board of Trade inspectors of railways. Colonel William McMurdo, a Crimean War veteran, was the honorary colonel of this body, which could undertake a range of railway works in wartime, whether of construction, demolition or reconstruction, and could supply military engineers with information, advice and the labour power of thousands of navvies. The corps was dedicated to ensuring that the railways acted in combination whenever the country was in danger, and to devising schemes in peacetime for bringing troops together from distant parts of the country and concentrating them within given areas in the shortest possible time. The corps also produced, and updated annually, reports on timetables and rolling stock that could facilitate the movement of troop trains.[43]

For mobilization plans in the event of home defence, the volunteers had to await the formation in 1873 of the army's intelligence department, a forerunner of the general staff that would be created in the Edwardian era. In the late 1870s Colonel Robert Home, a Royal Engineer who served in the department, began the planning for the defence of the United Kingdom. He regarded the volunteer units as essentially garrison troops, and, in the event of an emergency, envisaged their movement by rail to defend key coastal locations (including the movement of volunteer detachments from Derby to defend Plymouth). But his premature death in 1879 stifled further planning[44] until the appointment of Major-General (later General Sir) Henry Brackenbury as deputy quartermaster-general and head of the intelligence branch in 1886. Appalled by the War Office's lack of an up-to-date mobilization plan, Brackenbury criticized the inability of Britain to place two complete army corps in the field, either for foreign service or for home defence. He deprecated the absence of any organizational provision for the assembly, supply or transport of the scattered forces in the United Kingdom, especially since there were several completely undefended points on the coast, within an easy four days' march of London, that were suitable for an enemy landing. 'London', he observed, 'the richest town in the world, lies undefended at the mercy of the invader.'[45]

Having secured Treasury approval in October 1887 for a new Mobilisation and Home Defence section, Brackenbury urged its new head, Colonel (later Major-General Sir) John C. Ardagh, to return quickly from Egypt because the 'mobilization of our field army for home defence has yet to be worked out'.[46] In the subsequent planning Ardagh envisaged that regular forces (with some militia in the third corps) were to be stationed at railway junctions between London and the coast to give them maximum mobility by railway in the event of an invasion. While the 'judicious dispersion at railway stations' would permit 'an extraordinary facility of movement' for the infantry brigades, it would also bring together the allocated artillery in 'the proportion of two or three batteries per brigade'. Eighteen volunteer brigades, of six regiments apiece, were to fill in between the regulars and London. They were to assemble and undertake brigade and divisional manoeuvres (as well as the entraining and detraining of the different arms) in six entrenched camps at Aldershot, Caterham, Chatham, Tilbury, Warley and Epping, surrounding the capital.[47] In a subsequent paper entitled 'The Defence of London', Ardagh described how the volunteers would erect and man fixed fortifications around the capital. In an emergency, added Ardagh, the War Office expected assistance from the Railway and Engineer Volunteer Staff Corps in the distribution of administrative and executive staff and the arrangements for the supply of the necessary labour and materials, to ensure the construction of defensive works.[48]

These were all rather theoretical calculations but they coincided with broader concerns in contemporary military planning. In the short term another invasion scare erupted in the summer of 1888,[49] and the adjutant-general, Viscount Wolseley, intervened dramatically in a debate within the House of Lords to claim that 'our defences at home and abroad are at this moment in an unsatisfactory condition, and that our military forces are not organized or equipped as they should be to guarantee even the safety of the capital'.[50] Naval propagandists rallied in opposition to the 'Bolt from the Blue' thesis, with Vice-Admiral Philip Colomb, an eminent naval historian, and his brother, Captain Sir John C. R. Colomb, MP, asserting that the Royal Navy retained absolute primacy in home defence, despite the shortcomings of the fleet revealed in the naval manoeuvres of 1888. Embarrassed by this public dispute between the armed services, and the evidence of the chronic incapacity of the Royal Navy, the government of Lord Salisbury resolved the dispute in favour of the Senior Service. Under the provisions of the National Defence Bill of 1889 it proposed a massive expansion of the naval building programme. Meanwhile the army had to content itself with a memorandum drafted by the secretary of state for war, Edward Stanhope, which ranked the retention of two army corps at home for home defence as fourth in its order of priorities but ahead of the 'improbable' employment of one of those corps 'in a European war'.[51]

Channel Tunnel

Nevertheless, the military authorities had the satisfaction of thwarting the one perceived threat from a rail-based invasion through a projected Channel Tunnel. Known as the submarine railway, it had the backing of railway enthusiasts on both sides of the Channel, and the Channel Tunnel Company was formally incorporated and registered in London in 1872. Although the project stalled on the British side through a lack of funding, Sir Edwin Watkin, the Liberal MP and chairman of the South Eastern Railway Company, revived the scheme by sinking shafts in his company's property near Dover. By June 1881, he announced that experiments had succeeded, and that he had reached understandings with the French promoters.[52]

Within months a public outcry erupted as admirals, generals, newspaper editors and novelists denounced the prospect of a submarine railway linking England with the Continent. James Knowles, editor of the prestigious journal, *The Nineteenth Century*, organized a mass petition against the 'military dangers and liabilities' of the project.[53] While the Board of Trade remained mindful of the possible commercial benefits, officials debated the security issues before an inter-departmental committee without reaching a consensus. Hugh Childers, the secretary of state for war in the cabinet of William Ewart Gladstone, seized the initiative and appointed a War Office committee under Major-General Sir Archibald Alison to review all the security issues involved. On 12 May 1882 it issued a critical report, concluding that all the proposed measures for rendering the tunnel impassable to an enemy could not meet 'every imaginable contingency'.[54]

Wolseley, the duke of Cambridge, then the field marshal commanding-in-chief, and General (later Field Marshal) Sir J. L. A. Simmons led the military criticism.[55] The submarine railway posed 'a danger to the national existence of England', argued Wolseley: whoever held the Calais end could seize the Dover end by a *coup de main* and then send a large army through the tunnel. Seizing the tunnel, he claimed, by 'a few thousand men coming through a tunnel in the middle of the night in three or four trains' would be 'a very simple operation, provided it was done without any previous warning or intimation'.[56] The duke shared these fears, citing the precedent of the Fenian attempt to seize Chester castle in 1867. He doubted that anyone could guarantee the security of Dover castle once the tunnel was built. Like General Simmons, he feared that as soon as Britain became linked to the

Continent, she would have to accept Continental risks and take out similar insurance, namely a large standing army based upon compulsory service.[57]

Such fears could not be ignored. On 20 April 1883 a joint parliamentary select committee was convened under the chairmanship of Lord Lansdowne.[58] Having reviewed all the various reports and memoranda, it issued a draft report on 10 July 1883, which cast doubt upon most of the theories about how the tunnel could be seized in peace or war, particularly the notion of a *coup de main* in peacetime. It also accepted that there were numerous ways in which a tunnel could be rendered impassable in an emergency, yet it agreed with the conclusion of the Alison committee that Britain could not rely upon even the most 'comprehensive and complete arrangements' to deny the tunnel to an enemy 'in every conceivable emergency'. It recognized, too, that it would be hugely expensive to construct and man a first-class fortress to defend the Dover exit and to reorganize and re-equip the forces available for home defence. Finally, it accepted that only a large standing army, based upon compulsory service, could guarantee the nation's safety should the tunnel exist. Unwilling to support this 'gigantic evil', a majority of the committee advised that parliament should not approve the project, a conclusion that effectively proved 'the death blow of the Channel Tunnel scheme'.[59]

So railways had caught the military imagination. They had figured prominently in the notable invasion alarms, many of which were stimulated by fears triggered as much by technological developments as by threats from neighbouring powers. The railways offered the tantalizing prospect of deploying and concentrating defensive forces rapidly, although nearly all these schemes highlighted deficiencies in the numbers, organization and equipment of the forces retained at home. Given the predominance of the Royal Navy in the Channel and North Sea, these alarms had limited credence, attracted less support and faded relatively quickly. Yet the capacity of railways to act as a 'force multiplier' was not lost on the Victorian army as it grappled with military operations in Europe and across the empire.

Notes

1 Parl. Deb., Third Series, vol. 82, col. 1224 (30 July 1845).

2 As described by Richard Cobden, a radical politician, apostle of free trade and campaigner for peace and the reduction of the naval and military expenditures: R. Cobden, *The Three Panics: An Historical Episode* (London: Ward & Co., 1862).

3 PP, *Second Report from the Select Committee on Army and Ordnance Expenditure*, no. 499 (1849), IX, p. xlviii; M. S. Partridge, *Military Planning for the Defense of the United Kingdom, 1814–1870* (New York: Greenwood Press, 1989), pp. 70–1.

4 BL, Add. MSS 40461, fos. 215–16, Peel MSS, Duke of Wellington, 'Memorandum on the Means of Defence of the Country', 10 September 1845.

5 Ibid. fos. 341–6, Peel MSS, Sir G. Murray to Wellington, 8 November 1845.

6 J. Luvaas, *The Education of an Army: British Military Thought, 1815–1940* (London: Cassell, 1964), p. 75; see also Lt.-Col. The Hon. G. Wrottesley, *Life and Correspondence of Field Marshal Sir John Burgoyne, Bart.*, 2 vols. (London: Richard Bentley & Son, 1873), vol. 1, pp. 427–8.

7 TNA, WO 80/9, Murray MSS, Sir J. Burgoyne, 'Observations on (some of) the probable effects of the Modern Application of Steam Power in operation of attack & defence', 11 September 1845; see also Partridge, *Military Planning*, p. 80.

8 See his correspondence in respect of various railway proposals in 1844–47, TNA, WO 44/530.

9 TNA, WO 44/123, 'Projected Railroad from London to Dover', Col. J. R. Arnold to Inspector General of Fortifications, 23 January 1836, and minute by General Sir F. Mulcaster.

10 In 1781 the French had invaded Jersey and local militia had fought alongside regular regiments in the battle of Jersey. Fifty years later in 1831, in honour of its services in that battle, the militia was renamed as the Royal Jersey Militia.

11 TNA, WO 44/77, 'The Memorial of Jersey Railway and Pier Company', 23 February 1846, and Colonel Dixon's report, 'From the Commanding Royal Engineer Jersey to the Inspector General of Fortifications on a Memorial from the Railway and Pier Company Jersey to the Board of Ordnance', 10 March 1846.

12 Wrottesley, *Life and Correspondence of Burgoyne*, vol. 1, p. 408.

13 There were several different gauges on the principal railway lines of Victorian Britain in the mid-1840s. The Gauge Act (1846) ruled in favour of the standard gauge of 4 feet 8½ inches (143.5 cm) with certain exceptions, but it was not until 1892 that all gauges were standardized. J. Simmons and G. Biddle (eds.), *The Oxford Companion to British Railway History from 1603 to the 1990s* (Oxford: Oxford University Press, 1997), pp. 523–4.

14 PP, Sir John Fox Burgoyne (Qs. 6346, 6348), Minutes of Evidence before the Gauge Commissioners, no. 34 (1846), XVI, pp. 328–9.

15 PP, *Report of the Gauge Commissioners*, no. 34 (1846), XVI, p. 6, and Sir J. W. Gordon (Q. 6328) and Sir J. Burgoyne (Qs. 6347 and 6353), Minutes of Evidence, pp. 326, 328–9 and 330.

16 This paper was produced on 7 November 1846 and later published under the title 'Observations on the Possible Results of a War with France, under Our Present System of Military Preparation' in Capt. The Hon. G. Wrottesley (ed.), *The Military Opinions of General Sir John Fox Burgoyne, Bart.* (London: Richard Bentley, 1859), pp. 1–23.

17 'Report on the Defence of the Country, Submitted to the Cabinet by Lord Palmerston', 17 December 1847, reproduced in Wrottesley, *Life and Correspondence of Burgoyne*, vol. 1, pp. 436–44.

18 BL, Add. MSS 43750, fo. 23, Broughton diaries, Cabinet, 24 April 1847, and Add. MSS 43751, fo. 62, Cabinet, 20 December 1847.

19 Duke of Wellington to Burgoyne, 9 January 1847, reproduced in Wrottesley, *Life and Correspondence of Burgoyne*, vol. 1, pp. 444–51, at p. 445.

20 Longford, *Wellington*, pp. 460–2. There may also have been a man named Pigou involved' see Luvaas, *Education of an Army*, p. 76.

21 'The National Defences', *The Times*, 1 December 1847, p. 12; 'The National Defences', *Morning Chronicle*, 6 January 1848, p. 4; see also 'The Duke of Wellington upon the National Defences', *Bradford Observer; and Halifax, Huddersfield and Keighley Reporter*, 6 January 1848, p. 4; editorial, *Sheffield and Rotherham Independent*, 8 January 1848, p. 8; Spiers, *Army and Society*, pp. 91–2; I. F. W. Beckett, *The Amateur Military Tradition 1558–1945* (Manchester: Manchester University Press, 1991), pp. 146–7.

22 For a full account of the debates leading up to the Chobham camp of 1853, see Strachan, *Reform of the British Army*, pp. 161–6.

23 Royal Archives (RA), VIC/MAIN/E/42/13, Prince Albert MSS, Horse Guards, Memorandum, 17 November 1847.
24 For example, RA, VIC/MAIN/E/42/16, Secretary of the London and South Western Railway to General Sir W. Gordon, November 1847 (sent by the duke of Wellington to Prince Albert), and VIC/MAIN/E/42/20, Wellington and South Western Railway to General Gordon, December 1847.
25 RA, VIC/MAIN/E/44/23a, Prince Albert, memorandum sent to Lord Hardinge, 10 November 1852.
26 Strachan, *Reform of the British Army*, p. 166.
27 W. Malins, *A Plan for Additional National Defences* (London: J. Ridgway, 1848), pp. 9, 29–30, 32.
28 J. Anderson, *System of National Defence* (Edinburgh, 1853), pp. 13–17.
29 Beckett, 'Britain', p. 24, and Beckett, *Amateur Military Tradition*, pp. 148–9; see also Spiers, *Army and Society*, p. 92.
30 Sir J. Burgoyne, 'Remarks on the Military Condition of Great Britain' (1850) in Wrottesley (ed.), *Military Opinions of Burgoyne*, pp. 24–61, at p. 56.
31 TNA, WO 44/534, Sir J. Burgoyne, 'Memorandum on Dorchester and Exeter Proposal', 23 April 1852; see also Simmons, *Victorian Railway*, p. 366.
32 This is reproduced in full as 'Militia and Volunteers' in Wrottesley (ed.), *Military Opinions of Burgoyne*, pp. 91–112, at pp. 110–12; see also Wrottesley, *Life and Correspondence of Burgoyne*, vol. 1, p. 500 note.
33 While there is now a wide range of names for the insurrection in India, 'mutiny' is preferred here, as it is in the best military history, S. David, *The Indian Mutiny* (London: Viking, 2002). Although the event began with a sepoy mutiny in Meerut before spreading to other regions, such as Oudh, and attracted several Maratha leaders (notably Lakshmibai and the Rani of Jhansi), it destabilized neither the Bombay presidency nor the Madras presidency. Several princely states – Mysore, Travancore, Kashmir, Hyderabad and Rajputna – remained unaffected, while many sepoys as well as Sikhs from the Punjab and the Gurkhas fought for the British.
34 There had been offers to form rifle clubs in 1852 and two had survived, the Exeter and South Devon Corps and the Victoria Rifle Corps in London: Beckett, 'Britain', p. 24; H. Cunningham, *The Volunteer Force: A Social and Political History 1859–1908* (London: Croom Helm, 1975), pp. 6–7, 9–10; Burgoyne, 'Cherbourg' in Wrottesley (ed.), *Military Opinions of Burgoyne*, pp. 61–4.
35 Cunningham, *Volunteer Force*, pp. 11–12.
36 Though mainly in rifle corps, they also included artillery, engineer, light horse and mounted rifle corps: I. F. W. Beckett, *Riflemen Form: A Study of the Rifle Volunteer Movement, 1859–1908* (Aldershot: Ogilby Trusts, 1982), pp. 61, 63–4, 69–75, 128; Cunningham, *Volunteer Force*, p. 12.
37 Raised in December 1859, and renamed as the 'Railway Rifles' from 1861, the corps was converted in 1880 into the 11th Middlesex Rifle Volunteer Corps or the 11 (Railway) Middlesex Rifle Volunteer Corps.
38 P. W. Kingsford, *Victorian Railwaymen: The Emergence and Growth of Railway Labour 1830–1870* (London: Cass, 1970), pp. 76–7.
39 Beckett, 'Britain', pp. 36–7.
40 Beckett, *Riflemen Form*, p. 98. Macdonald reported that a similar crowd gathered to greet the returning volunteers but appeared at the wrong station, so leaving the volunteers to arrive at another station 'where there was not a soul to meet them'! J. H. A. Macdonald, *Fifty Years Of It: The Experiences and Struggles of a Volunteer of 1859* (Edinburgh: William Blackwood, 1909), pp. 161, 202.
41 D. Howie, *History of the 1st Lanark Volunteers* (Glasgow: David Robertson, 1887), pp. 133–43.
42 W. H. Chaloner, *The Social and Economic Development of Crewe, 1780–1923* (Manchester: Manchester University Press, 1950), p. 273.
43 E. A. Pratt, *The Rise of Rail-Power in War and Conquest 1833–1914* (London: P. S. King & Son, 1915), pp. 179–83; for a narrative history of the corps, see Maj.

C. E. C. Townsend, *All Rank and No File: A History of the Engineer and Railway Staff Corps, RE, 1865–1965* (London: Engineer & Railway Staff Corps RE (TAVR), 1965).

44 B. Bond, *The Victorian Army and the Staff College, 1854–1914* (London: Eyre Methuen, 1972), pp. 120–1.

45 TNA, WO 33/46, Maj.-Gen. H. Brackenbury, 'General Sketch of the Situation Abroad and at Home from a Military Standpoint', 3 August 1886; see also E. M. Spiers, *The Late Victorian Army 1868–1902* (Manchester: Manchester University Press, 1992), p. 226.

46 TNA, PRO 30/40/2, Ardagh MSS, Brackenbury to Ardagh, 13 October 1887.

47 TNA, PRO 30/40/13, Ardagh MSS, 'Defence of England', 17 April 1888.

48 Ibid. 'The Defence of London', 16 July 1888; see also T. G. Fergusson, *British Military Intelligence, 1870–1914: The Development of a Modern Intelligence Organization* (Frederick, Maryland: University Publications of America, Inc., 1984), pp. 84–5.

49 A leading military theorist and champion of the volunteers, Lieutenant-General Sir Edward Hamley, MP, was one of the alarmists on this occasion: E. Hamley, 'The Defencelessness of London', *Nineteenth Century*, vol. 23, no. 135 (May 1888), pp. 633–40; see also J. Gooch, *The Prospect of War: Studies in British Defence Policy, 1847–1942* (London: Frank Cass, 1981), pp. 7–8.

51 Spiers, *Victorian Army*, pp. 228–9 and appendix 3, 'The Stanhope Memorandum', p. 337; see also I. F. W. Beckett, 'The Stanhope Memorandum of 1888: A Reinterpretation', *Bulletin of the Institute of Historical Research*, vol. 57, no. 136 (1984), pp. 240–7.

52 K. M. Wilson, *Channel Tunnel Visions 1850–1945: Dreams and Nightmares* (London: Hambledon Press, 1994), pp. 11, 19–21.

53 I. F. Clarke, *Voices Prophesying War 1763–1984* (Oxford: Oxford University Press, 1966), pp. 110–11.

54 TNA, WO 33/39 and WO 32/5299, 'Précis of the Channel Tunnel Negotiations'; Wilson, *Channel Tunnel Visions*, pp. 22–36.

55 Sir John Lintorn Arabin Simmons (1821–1903) was a Royal Engineer who held appointments as inspector of railways in 1847 and later as secretary of the Railways Commission in 1850. He investigated many railway accidents, including the Dee Bridge disaster of 1847. Major-General Sir John Adye, surveyor-general of the ordnance, and Colonel Sir Andrew Clarke, commandant of the School of Military Engineering, were among a minority of senior officers who reckoned that the security issues posed by a tunnel could be managed effectively: see Wilson, *Channel Tunnel Visions*, pp. 26–7, 33–5, 37.

56 PP, *Correspondence with Reference to the Proposed Construction of a CHANNEL TUNNEL*, C. 3358 (1882), LIII, 'Memorandum by Sir Garnet Wolseley', 16 June 1882, pp. 272, 282, 284–5, 297.

57 Ibid. 'Observations by His Royal Highness the Field-Marshal, Commanding in Chief', 23 June 1882, pp. 303–4; Sir J. L. A. Simmons, 'The Channel Tunnel', *Nineteenth Century*, vol. 11 (1882), pp. 663–7.

58 Henry Petty-Fitzmaurice (1845–1921) was the fifth marquess of Lansdowne, and later held the posts of governor-general of Canada (1883–88), viceroy and governor-general of India (1888–94), secretary of state for war (1895–1900) and foreign secretary (1900–5).

59 TNA, WO 32/5299, 'Précis of the Channel Tunnel Negotiations'; PP, *Report from the Joint Select Committee of the House of Lords and the House of Commons on the Channel Tunnel; together with the Proceedings of the Committee, Minutes of Evidence, and Appendix*, C. 248 (1883), XII, pp. xx–xxvi, xxxv–xxxviii, xliv–xlv.

3

Railway experiments in mid-Victorian wars

However useful railways were in aid of the civil power, and however great a potential asset in defence of the United Kingdom, the army would appreciate them fully only once they had been utilized in war. Over nearly thirty years from the Crimean War (1854–56) to the intervention in Egypt (1882), the Victorian army would both experiment with railways in some wars and observe the employment of railways by other armies in foreign wars. Although the army found itself engaged in colonial wars in every year from 1857 to 1902 other than 1869 and 1883,[1] and its officers and men benefited, in Wolseley's words, from 'the varied experience, and frequent practice in war',[2] it could not employ railways systematically when forced to fight over remote and barely accessible terrain, sometimes in inhospitable climates, and usually in conflicts where there was every incentive to win decisively and in a minimum of time.[3] Nevertheless, an interest in exploiting railways would emerge, stimulated by Britain's pioneering use of railways in the Crimean War, observations of foreign wars and limited colonial experiments.

The Crimean War and its immediate aftermath

The Balaclava Railway, sometimes grandiloquently described as the Grand Crimean Central Railway, fascinated contemporaries, with many of its peculiar features attracting the commentary and plaudits of railway historians.[4] The construction and use of a railway had not been anticipated when Britain and France declared war on Russia on 28 March 1854 in support of the Ottoman empire. It was only after the early engagements proved inconclusive, including the abortive Russian assault at the battle of Inkerman (5 November 1854), and the allies became involved in a protracted siege of Sevastopol, that logistic issues assumed a major prominence over the forthcoming winter. Operating out of a narrow harbour at Balaclava some thirteen kilometres south of Sevastopol, the British had to bring supplies and ammunition up a road that was barely a track, via the village of Kadikoi about three kilometres from Balaclava. The road then wound its way up a steep climb via the Col of Balaclava to the bleak Khersonese Uplands, about 183 m above sea level, where the British forces were encamped and trenches were dug before the Russian fortress. The transport, commissariat and medical services were already struggling to cope before a ferocious storm,[5] accompanied by sleet and snow, on 14 November 1854 devastated the British encampments and wrecked the shipping outside Balaclava harbour. Twenty-four ships foundered, including the steamship *Prince*, full of winter clothing. As the British army, initially some 22,000 strong,[6] was already ravaged by cholera and sickness, and was now dependent upon supplies moving across a muddy track, the strains upon the men, some of whom were working in bitter conditions within mud-filled trenches, became increasingly apparent.

Bleak descriptions filled private correspondence; William Govett Romaine, the deputy judge advocate to the Army of the East, recorded that barely 16,000 rank and file were 'under arms' by 18 November, and, a fortnight later, that a recent ten-day mortality rate had equalled the losses at the battle of Alma (20 September 1854): 'We have had a black time and the Ravens of the army have been croaking desperately.'[7] The British press had published further revelations in uncensored letters from serving officers and other ranks, while *The Times*, then boasting an unrivalled circulation of 40,000 readers, published the uninhibited, graphic and often caustic commentary of its own correspondent, William Howard Russell. A fierce critic of military mismanagement, Russell had lauded the martial valour of soldiers in earlier battles but now bitterly depicted their sufferings as tents and trenches filled with water, roads became impassable with mud, 'and not a soul seems to care for their comfort or even for their lives'.[8] Appalled by such reports, Samuel Morton Peto, a Whig MP and one of the leading railway contractors, persuaded his partners, Edward Betts and Thomas Beatty, that they should offer to build a railway from the port of Balaclava to the forces outside Sevastopol. The government accepted their offer to build a working railroad within three weeks of landing at Balaclava, at an initial cost of £45,000 for the provision of stores, plant and equipment and £55,000 for the freight charges to the Crimea. However, the freight charges alone soared to £135,000 with another £14,000 in salaries, outfits and provisions for the men who arrived at the beginning of February 1855.[9]

Donald Campbell, a surveyor, had already managed to find a possible route (see Map 2), following the only road, which included a gradient of 1 in 14. A stationary engine would have to pull the railway wagons up a steep incline before horses could haul them on to Lord Raglan's headquarters. Beatty with an advance party of 500 men arrived on 19 January 1855 to begin the preparations in Balaclava, where he encountered scepticism from several senior officers. Lord Raglan wondered if the 'train road' could be 'laid by magic',[10] while Burgoyne reckoned that civil engineers could achieve 'great things' only through their access to 'an unlimited expenditure'.[11] Like Lieutenant-Colonel Anthony Sterling and Captain Henry Clifford, Burgoyne reckoned that a macadamized road was a higher priority than a railway.[12] Once the full complement of navvies arrived, Burgoyne even requested that they might be employed in making defensive positions for the soldiers, a 'bald proposal' that Beatty brusquely refused.[13] These military views have been characterized by railway historians as those of 'the old army', commanded by aristocrats and elderly veterans of the Peninsular War (1808–14), with scant evidence of 'military imagination' or appreciation of a railway 'organised and built by civilians'.[14]

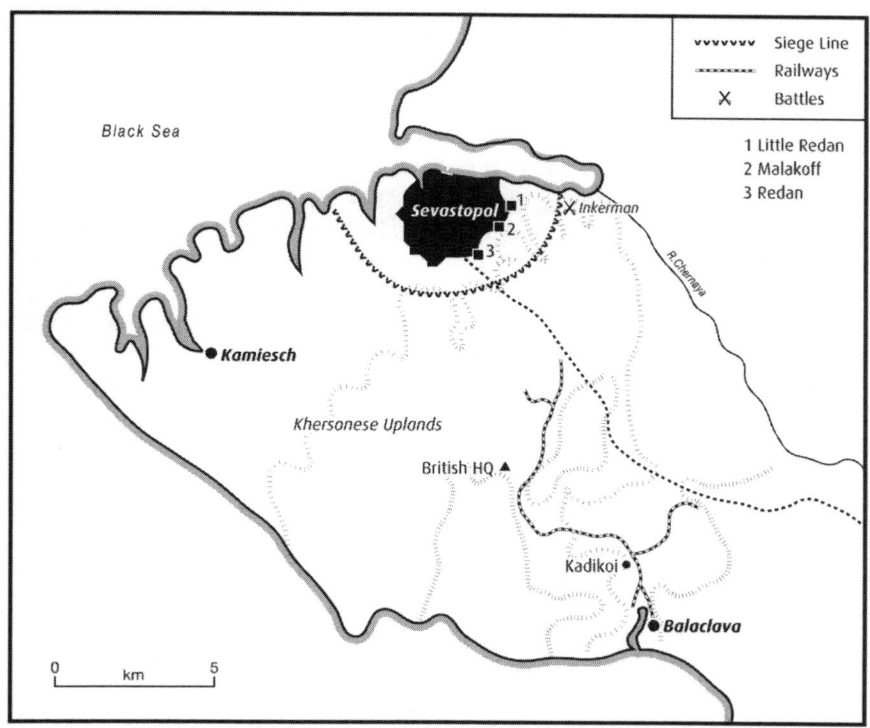

Map 2 The Balaclava Railway

Yet Burgoyne, as evidenced by his writings on home defence, appreciated fully the potential of railways, and the case for building a road was hardly misguided: indeed, a much better road was constructed in the following autumn. Military opinion was not universally hostile. Sir Colin Campbell, the commander of the Highland Brigade based in Kadikoi, expressed confidence in the potential of railways,[15] and Beatty received invaluable assistance from 150 (later 200) soldiers from the 39th (Dorsetshire) Regiment of Foot, as well as some 200 Croat workers, who were employed in carrying stones for the road, and another 200 sailors who were lent over a period of ten days.[16] With the remaining navvies and all their equipment arriving at the beginning of February, the laying of rails began on 8 February, and the first horse-drawn supplies reached Kadikoi on 23 February. Even military sceptics applauded the rate of progress,[17] and, by 25 February, Staff Assistant Surgeon George Lawson, in commenting on 'our railway', remarked that

> It has passed a little village called Kadikoi, outside this town (Balaklava) about a mile and a half, and is close to the Cavalry camp. That portion of the line is now in use, and stores and wooden huts etc. are being carried up in large quantities by it: the large railway carts being dragged along the line by some gigantic English carthorses.[18]

During the lull in hostilities caused by the wintry conditions, the construction of the railway, as represented in the illustrated and printed press, aroused widespread interest. Roger Fenton, the photographer, who had arrived in the Crimea in February 1855, took several photographs of the railway yard; William Simpson, the lithographer, painted a famous scene of the main street in Balaclava with the cart horses at work on the railway; and Constantin Guys prepared a major sketch of the railway for the *Illustrated London News* (24 March 1855). Russell observed how 'strange' it was that 'the great invention of recent days should be to facilitate the operations of war and to destroy life',[19] while Romaine was confident by 11 March that 'The Railway will turn out famously & be of capital consequence to our operations . . .'.[20] The fact that the eleven-kilometre line took twice as long to build as Peto forecast, and did not reach headquarters until 26 March 1855, disturbed few observers. From its inception the railway carried 200 tons of material daily, was double-tracked to Kadikoi (see Figure 2) and enhanced the use of the Balaclava wharf.[21]

Following further surveys new lines extended and improved the railway, offsetting some of the risks taken by the hasty construction and later providing more sidings, passing loops along the single track and a branch line for the Sardinian allies. Ultimately the railway ranged

Figure 2 'Progress of the Balaclava Railway to the Church of Kadikoi', *Illustrated London News*, vol. 26, no. 753 (31 March 1855), p. 272

over 22.5 km, with two and then three stationary engines and the first of five locomotive engines in action from 8 November onwards.[22] Operated initially by civilians and the navvies of the newly formed Army Works Corps, this arrangement soon lapsed as the civilians returned home after the expiry of their six-month contract. Subsequently Royal Engineers drove the steam locomotives at sea level from Balaclava, taking forty-eight minutes to travel the mile and a half, but as none of these engines could manage the gradient of 1 in 25, the stationary engine had to haul the wagons up the incline to Kadikoi. Thereafter the newly formed Land Transport Corps, under Colonel William McMurdo (and later Colonel Edward R. Wetherall), operated the railway by using horsepower to drag the wagons up the further incline and thence onto the upland campsite. The return of the wagons to Balaclava was primarily by gravity, a risky undertaking that caused numerous accidents. Nevertheless, by the late autumn of 1855 the railway, as Russell observed, had acquired 'an appearance of solidity and permanence', and subsequently brought 'the familiar sounds of Wolverhampton or of Didcot'.[23]

Though cumbersome, and at times dangerous, the railway proved itself better as a means of supplying British forces than the cart roads, which still carried supplies up to the British forces. The initial construction was also timely, enabling the movement of newly arrived guns, mortar shells and other ordnance for the 'Second Bombardment', launched on 4 April 1855. The firing of 47,000 rounds over ten days was unprecedented in intensity, and it was followed by even heavier bombardments in the following June, and then in August, when 307 guns fired 150,000 rounds in four days. Although the railway supported all these operations, it hardly merited the accolade 'the railway that won a war'.[24] None of these bombardments proved decisive, and, despite the central deployment of the British gunners, their contribution to the allied bombardment of Sevastopol was always subordinate to that of the French. While the British failed in their assaults on the Great Redan (18 June and 8 September 1855), and took no part in battle of Chernaya (16 August 1855), the French stormed the Malakoff redoubt and the Little Redan (8–9 September 1855), forcing the Russians to abandon the southern side of Sevastopol.[25]

If the railway did not win the war, it still performed a significant role in the Crimea. It converted many of the military sceptics, not least by relieving the infantry of some of the burden of 'shot carrying' from the wharf up to the front line.[26] Coupled with transport on the new road, it helped to sustain the British and the recently arrived Sardinian allies through the second winter. The railway probably saved on the wastage of horses and mules, although the motley personnel of the Land Transport Corps (inexperienced, untrained and ill-supervised British drivers outnumbered by an even worse crew of Spaniards, Maltese, Croats, Tartars and Turks with barely one officer for every 250 men) had an appalling reputation for the care of their animals, with over a thousand dying or being destroyed in November 1855.[27]

At its peak the railway (operated by a thousand men) carried 700 tons daily, and during the second winter, which was less severe than its predecessor, the railway sustained a daily rate of 138 wagon-loads, transporting 414 tons of supplies, huts, food, fuel and other requirements for the British and Sardinian forces. The railway also removed the sick and injured from the front line and conveyed them directly to hospital ships in Balaclava. On visiting the Crimea in July 1855, the former secretary of state for war, Henry Pelham, the fifth duke of Newcastle, was hugely impressed by the 'cleanliness of the harbour', and the fact that the 'railroad comes down to where the hospital ships lie, so that the sick can be carried straight on board'.[28] Finally, in the wake of the armistice, the railway proved immensely helpful in removing a vast amount of *matériel* from the peninsula. Russell, an astute commentator to the last, noted on 11 April 1856 how 'Each division collects about 4,000 shot a-day, and they are carried to Balaklava as fast as the means at our disposal – railway and land transport – will permit.'[29]

At the end of the conflict Burgoyne sent a letter to Peto, handsomely admitting that it was impossible to overrate the services of the railway or its effect on shortening the time of the siege and alleviating the sufferings of the troops.[30] The railway was also part of a concerted effort to improve the supplies and transport during the war, prompting some railway historians to conclude that it taught the 'more far-seeing elements' in the army to recognize the importance of logistics in Victorian warfare: the Land Transport Corps ultimately became the Royal Corps of Transport, and the Army Works Corps, the Pioneer Corps.[31] Drawing such links over a period in excess of 100 years is somewhat fanciful, as the shortcomings of the new bodies were all too apparent towards the end of the war. General Sir James Simpson, who assumed command of the army after Raglan's death on 28 June 1855, denounced the Army Works Corps as 'by far the worst lot of men ever yet to be sent here', while General Sir William Codrington, who succeeded Simpson as commander in the following November, criticized the limited capacities both of the locomotives and of the railway carriages sent to the Crimea, and condemned the whole structure of the Land Transport Corps. A 'total change of system', he asserted, 'must take place if the army is to be made moveable'.[32] Renamed as the Military Train within a year, the body would undergo major reforms before emerging as the Army Service Corps a decade later. The Crimean railway meanwhile had proved an exceptional enterprise: it had been possible only because of the proximity of the conflict, the seaborne access and the lull in siege hostilities during the winter months. None of these factors would apply in Britain's next major war: the Indian Mutiny (1857–59).

Railways and the Indian Mutiny

The revolt of sepoys at Meerut (10 May 1857) and their capture of Delhi, the old imperial capital, on the following day posed an unprecedented challenge to British rule in India. Only a firm and decisive response could stem the consequences, as mutinies and massacres of Europeans and Indian Christians spread through Oudh, central India and parts of the north-west provinces, including Rajputana and the Punjab, and as sieges of British garrisons began at Cawnpore (Kanpur) and the residency at Lucknow. Faced with 311,000 Hindu, Muslim and Sikh troops, some of whom held back initially (though many, including Gurkhas and Sikhs, later supported the British in suppressing the rebels),[33] the authorities in Calcutta could only call upon 40,000 British soldiers in India, namely those serving in the Queen's regiments, the all-British East India Company units and the white officers of Indian regiments. While there were British forces in the Punjab, there were only two Queen's regiments between Calcutta and Meerut (the 32nd at Lucknow and the 10th at Dinapore near Patna). Given the unfolding nature of the crisis, and the need to reassure Indian peoples that the British would prevail,[34] the authorities had to act quickly. They received the assistance of some battalions, which were returning from the Persian War (1856–57), and sought the diversion of units from an expeditionary force bound for China, but they could neither wait upon reinforcements from the United Kingdom nor construct additional means of transportation.

By May 1857 there were only three railway lines in India: the Great Indian Peninsula Railway from Bombay to Poona via Thana; the Madras Railway running to Arcot; and the East Indian Railway, a 120-mile track running from Howrah near Calcutta to Raniganj. These were all commercial ventures, supported by funding from the Indian government and guaranteed up to 5 per cent interest on the capital expended in exchange for a measure of subordination, not least in respect of route placement. Underpinned by the famous Railway Minute of the then governor-general of India, Lord Dalhousie (23 April 1853), the trunk lines reflected his desire to exploit the political, social, commercial and potentially military benefits of creating a network of railways linking all three Presidency ports, Calcutta, Madras and Bombay. They had generally proved popular in India, apart from a clash in 1856 between Santhal tribesmen, armed with bows and arrows, and the engineers of the East Indian Railway.[35] That railway would prove to be the only one significantly involved in the mutiny as reinforcements arrived in Calcutta, ready to move into the hinterland. Among those reinforcements were the 900 men of the 1st Madras European Fusiliers, who arrived on 23 May 1857. They were commanded by Colonel James Neill, a dour Scottish Presbyterian, bent on fulfilling God's work and punishing the mutinous sepoys and anyone associated with them. Having sent 130 men up the Ganges River by steamer, Neill planned to embark another 100 by train, but the party arrived late on account of a squall. While the stationmaster threatened to despatch the train without them, other railway officials informed Neill that he might command a regiment but he 'did not command them'. Neill promptly placed the engineer and stoker under guard while ensuring that all his men boarded the train for a departure only ten minutes late. Neill then famously reprimanded the railway officials, describing their conduct as 'that of traitors and rebels' and assuring them how 'fortunate it was for them that *I* had not to deal with them'. Thereafter, noted Neill, 'the railway people are now most painfully civil and polite'.[36]

While Neill's ruthlessness earned plaudits from the governor-general downwards,[37] the railway played only a minor part in supporting the subsequent operations that were conducted hundreds of miles from its terminus at Raniganj. If the Mutiny confirmed that a rail network was a vital requirement in the maintenance of internal order, not least the planned line from Howrah via Raniganj and Allahabad to Delhi, the railway could not be extended at this time. In fact, the more exposed sections of track and facilities came under attack. In June, supporters of the Maulvi Liaquat Alia, who had proclaimed a jihad against the Christians, destroyed a station, a locomotive and carriage shops in the Allahabad district. At Barawi, some 37 km distant, railway staff and their families took refuge in a dry water tanks and held off attacks from rebels for over thirty hours before relief arrived from Allahabad. Even more dramatically an engineer, Richard Vicars Boyle, fortified a small building, later known as 'the Little House at Arrah'. When it subsequently came under attack from mutinous sepoys from the Dinapore cantonment, he held it for over a week, with the assistance of fifty Sikh police and fifteen railwaymen, before relief arrived from group of fusiliers and twelve men from the East Indian Railway Volunteer Force.[38] Ultimately the Mutiny had little effect upon the railway beyond retarding the extension of the East Indian network for six months, while the building of railways in the south and west proceeded 'in the face of natural rather than manmade obstacles'.[39]

Commentary on the use of railways in foreign wars

The British military was not preoccupied with its own experience in small colonial warfare. As major wars erupted in Europe and North America, it sent officers to observe their strategic and tactical developments, and/or hosted public meetings to reflect upon the 'lessons' of those conflicts. The Italian Campaign of 1859 (once known as the Franco-Austrian War and now as the Second War of Italian Independence) aroused widespread interest because of the popular support for Italian unification. In this instance a covert deal between Napoleon III and Camillo Benso, conte di Cavour, the prime minister of France's Crimean ally, the kingdom of Sardinia (better known as the Piedmont), advanced the cause. Under the secret treaty, France would support the Piedmont if it were attacked by Austria-Hungary in defence of its colonies, Lombardy and Venetia (and thereafter the Piedmont would cede Savoy and Nice to France). Following Sardinian military manoeuvres, which triggered an Austrian ultimatum, the ten-week campaign used railways much more extensively than ever before. For British commentators, railways enabled these adversaries to bring hundreds of thousands of men,[40] material and horses into conflict with unprecedented rapidity, and allowed them to conduct a war over a couple of months (from 29 April to 11 July). In a lecture before the Royal United Service Institution (RUSI) Major Frederick Miller, VC, RA, emphasized that railways in this war had

> assisted the ordinary means of locomotion hitherto employed by armies. By them, thousands of men were carried daily through France to Toulon, Marseilles, or the foot of Mont Cenis; by them troops were hastened up to the very fields of battle; and by them, injured men were brought swiftly back to the hospitals, still groaning in the first agony of their wounds.
>
> Moreover, the railway cuttings, embankments, and bridges presented features of importance, equal to, or superior, to the ordinary accidents of ground; and the possession of such features was hotly contested.[41]

While Miller appreciated that Napoleon III had relied upon French railways to convey his five corps and the Imperial Guard into action as quickly as possible,[42] he knew less about the logistical problems which left soldiers without food, and horses without fodder, at crucial times, and ultimately thwarted any attempt to follow up the allied victory at Solferino (24 June 1859).[43] Tactically, Miller grasped that the destruction of railway lines and bridges had become another distinctive feature of a campaign fought across a river-laced topography, and that the botched destruction of a railway bridge facilitated the Franco-Sardinian victory at Magenta (4 June 1859).[44] In a subsequent RUSI lecture delivered on 13 May 1864, Captain H. W. Tyler, RE, commented more fully on the Austrian use of railways to move troops and supplies from Verona northwards by the valley of Adige, eastwards to Venice and Udine, south to Mantua and westwards to Milan, Novara and Vercelli. He also described how the Austrians had tried to thwart enemy movements by tearing up rails and blowing up railway bridges, although such damage had proved easily repairable. More significantly the Austrians, he argued, had used the railway as a means of defence during the retreat from Magenta (as well as a method of extracting their wounded), while enabling the French to move soldiers and supplies rapidly to the front before removing wounded and prisoners to the rear.[45] What Tyler failed to appreciate were the physical difficulties that had bedevilled rail reinforcements from Austria (the heavy gradients of the Semmering Pass and the gaps in the line between Innsbruck and Bozen); the lack of rail organization which resulted in delays, blockages, congested stations and deficiencies in rolling stock in Vienna; and the staffing blunders that left trains stranded without instructions about their final destinations. Although the Austrians still moved army corps more quickly by rail than would have been possible if they had marched by road, they failed to reap significant benefits because of their hapless command and staff work. By contrast the French seemed to have exploited the railways more effectively, even augmenting their forces in battle by rail at Montebello (20 May 1859) and so fuelling the misguided impression that French had mastered this form of military movement.[46]

Captains Tyler and Luard, though, were much more impressed by the use of railways in the American Civil War, a war fought over several years (1861–65) across a vast terrain some 1,448 km long by 1,287.5 km broad.[47] Like Major Miller, they did not observe the military use of railways at first hand,[48] but realized from the printed commentary that railroads had had a transformative effect upon this war. American railways, by comparison with relatively indifferent American roads, were often of great length, especially in the North, and crossed numerous rivers spanned by wooden bridges and viaducts. During the Civil War they quickly became the principal means of moving and supplying large military forces, and so railway junctions became strategic objectives while rails, engines, bridges and rolling stock became objects of destruction. If railroads were more vulnerable than regular roads, they were also more easily repaired if the requisite men and materials were to hand. Tyler concluded that railways were a 'great advantage' in moving men and supplying them within a theatre; that even a single line, with a proper proportion of sidings, crossing places and rolling stock, could suffice, but a double track would be preferable; that railway junctions would often become strategic objectives; and that railways could be employed more advantageously on the defensive than in supporting offensive warfare. Writing before Major-General William Tecumseh Sherman's 'Atlanta Campaign', and the subsequent 'March to the Sea' (1 May to 1 September and 15 November to 21 December 1864), and poorly informed about developments in the western theatre, Tyler was convinced that the South had benefited from its interior lines of communication.[49]

Tyler had failed to realize that the South was not only suffering from shortages of men and *matériel* but had also failed throughout the war to manage the Southern railroads.[50] This failure of Confederate management found reflection in the failure to produce a single rail after 1861, and the lack of any planning directive or imposition on the southern railroads. It contrasted sharply with a managerial effort in the North, which enabled the Federal forces to surmount unforeseen, changing and ever-increasing challenges to meet their logistical requirements.[51] In his retrospective reflections on the war, Captain Luard grasped the significance of railway management. He quoted extensively from the report of Major-General Daniel McCullam upon the formation of a Federal Construction Corps, which originally numbered 400 men but expanded gradually to about 10,000. By forming 'skilled workmen in each department of railroad construction and repairs under competent engineers', responsibility could be delegated and work supported in diverse areas. Luard was understandably impressed by accounts of construction and repair in field conditions, such as the Chattahoochee bridge, 237.7 m long and 28 m high, reportedly 'completed by the construction corps' in four and a half days, and the repair within thirteen days of 57 km of track and 138.6 m of bridges, which General John B. Hood's army had destroyed in Georgia in October 1864. Luard acknowledged, too, that this railway organization had become so proficient after three years that it provided invaluable support for General Sherman's army when it drove through Georgia: 'the management of railroads', he concluded, 'is just as much a distinct profession as is that of the art of war, and should be so regulated'.[52]

Anglo-Abyssinian War, 1867–68

In a major African expedition, launched to rescue several missionaries and two envoys of the British government held captive by Emperor Tewodoos II (known as 'Theodore') of Ethiopia, the British laid another coastal railway. Faced with the immense logistical task of moving 13,000 British and Indian soldiers across 644 km of largely mountainous terrain, Lieutenant-General Sir Robert Napier was anxious about operating in 'a country of which we know so little', and was determined not to 'throw our troops beyond our means of feeding them'.[53] Accordingly, he sent an advance brigade to land in Annesley Bay on the Red Sea coast and construct a base camp at Zula, where his army and a vast support corps, ultimately comprising some 26,000 camp followers and 40,000 animals, including forty-four trained elephants, could assemble. The advance brigade brought rails and sleepers to lay a mile of tramway, and a dozen light trucks to operate a route from the high-water mark down to the newly constructed stone pier (see Figure 3). However, once it was decided that the advance into the Abyssinian Highlands would proceed from the Senafe Pass, Lieutenant Willans, assistant field engineer, undertook a survey for a tramway across the coastal plain from Zula to Kumayli at the entrance to the pass. Surveying the 19.3-km route commenced in November, and the construction works followed in December as rails and other materials arrived from Bombay.[54]

Yet the building process, supervised by several RE officers, and undertaken by men of the Army Works Corps, the 23rd Punjab Pioneers

Figure 3 William Simpson, 'The New Pier at Annesley Bay, 25 March 1868'

and the 2nd Bombay Grenadiers 'during the hot season',[55] proved excruciatingly slow. Having embarked on the expedition in late January 1868, Napier entered the interior before 19 February, when 9.7 km of track had been constructed. Only then was the land transport train able to start moving animals, stores and provisions by rail out to the six-mile siding.[56] So slow was the construction that it received far less coverage than its predecessor in the Crimea as most of the correspondents and special artists pressed ahead with Napier, while provoking a 'scathing' critique from G. A. Henty, the special correspondent of the *Standard*.[57] Henty assumed from the outset that the British soldiers like the 33rd (Duke of Wellington's) Regiment of Foot and the Indian forces had 'nothing' to fear in Abyssinia. 'It is not the enemy we fear', he argued, 'the enemy is contemptible; it is the distance, and the questions of provisions and transport.'[58] Absolutely correct in this forecast, he also affirmed that railway could prove of 'immense utility to the expedition'. Accordingly he despaired of the manner in which the railway crept forward across 'perfectly flat' terrain, with only little bridges required to cross the dry watercourses. As the construction was halted at the end of April, 1.6 km short of Kumayli, he ascribed this debacle to the failure to bring railway contractors with their gangers and platelayers from England, via Egypt. He described the single-line track as 'without exception, the roughest, most shaky, and most dangerous piece of railway ever laid down'.[59]

However perceptive these observations, they failed to explain the difficulties under which the Royal Engineers laboured. First, the Bombay authorities had shown scant appreciation of the requirements of the railway and gave priority to stores that would be required more urgently. They sent 'five different descriptions of rails . . . on four different principles of fixing',[60] and hardly aided disembarkation by stowing rails in the bottoms of ships, with carriages and engines above. They sometimes sent rails without spikes, and when the latter were discovered, the augurs (for boring holes in sleepers) were found on another ship. Two of the six engines were then pressed into service from mid-January, moving stores from the pier but not necessarily materials for the platelayers. Both the shipping supply and the process of disembarkation kept the construction operating on a 'hand to mouth' basis throughout.[61]

Staffing difficulties compounded these problems. Only half of the 1,200 labourers assigned to the railway ever worked on it, the platelayers proved inexperienced, and only one NCO could speak the language of the labourers. Some sepoys disliked the work, occasioning the odd flogging,[62] and they all struggled in the intense heat from February onwards. Allocated only one and half gallons of water each per day, the labourers barely worked from daylight to noon, with water brought out to them from Zula. Since the tank engines also consumed 1,000 gallons of water per day, the water demands stretched the capacities of the small condenser on the pier. Ultimately, these requirements were not met until three wells were sunk at 6, 11 and 14.5 km along the route, at depths of 15.2, 19.8 and 25.9 m respectively.[63] By the end of April, the construction process ground to a halt, with another siding 14.5 km from Zula and a loop at the terminus 1.6 km short of Kumayli. As the expedition was withdrawing from the hinterland, and 'the traffic on the line had now become so great, the commissariat department absorbed the whole of the rolling stock'.[64]

Fortunately the track, operating on a broad Indian gauge (5 feet 6 inches, or 1.7 m), was relatively straight, with few sharp curves but a gradual incline to a height of about 30.5 m above sea level. It then passed through a low range of hills along a river bank before descending to the Kumayli Plain. Some 19.4 km of track were laid, and they crossed eight iron girder bridges. The rolling stock, however, was hardly impressive. All of the four engines were light and not particularly powerful, with the best engine able to draw only fifteen small loaded trucks, none of which had springs or spring buffers. Their cast-iron axle bearings were soon worn through, leaving some trucks in need of repair after only a fortnight's use. Of the sixty wagons sent for use on the line, at least 40 per cent were always under repair, and it was only in May 1868 that wagons arrived with springs and spring buffers, but even their axles had to be adapted before they could be fitted with covers for passenger use. Nevertheless, by 21 March, the commissariat was working six trains a day, and by 24 May, it reported that it did not require any further carriage of stores from Zula to facilitate the withdrawal. As troops were now arriving at Kumayli from Magdala, they were conveyed by rail to Zula.[65] Over a period of three months the railway carried 9,000 tons of commissariat stores, 2,400 tons of materials, 14,000 troops and 10,000 followers, and 2,000 tons of baggage. This was done without accident and saved the troops and baggage animals from a trying march across the coastal plain during the hottest part of the year. 'The Abyssinian railway', in the opinion of Lieutenant Willans, 'was a great success' and a useful 'additional means of transport' at a cost of some £6,000 exclusive of rails and plant.[66] Captain George Tryon, RN, agreed: the railway, he declared, 'was nearly as necessary as anything that was taken in the Expedition, and very valuable'.[67]

Such enthusiasm was perhaps understandable from those most closely involved in the construction and use of the railway, but a tramway so short in length had made a very limited contribution to the withdrawal of the expedition. Napier had been acutely concerned about the state of his land transport,[68] and, during the withdrawal, had taken 'every precaution' to avoid the channels of deep passes 'after dark or when rain threatens' and had carried back 150 sick and wounded.[69] So a longer and more effective railway, built more quickly by experienced platelayers, might have proved even more useful, and several engineers argued that experienced civil engineers and their contractors should be solely responsible for laying and maintaining railways in future wars.[70] Even Willans, who remained an advocate of military construction, accepted that lessons had to be learned from Abyssinia, namely that a future military railway had to be based on a narrow gauge, with rails fixed to sleepers by devices known as chairs and flanged by fishplates (metal bars used to connect the ends of rails). As a military railway was bound to be rougher than a civil railway, all wagons should have springs and spring buffers.[71] Yet much of this was pure speculation: what had limited the scope, construction and utility of the Abyssinian railway was the decision as late as November 1867 to explore the route to the Kumayli Pass.[72] In a campaign that had to be completed before the onset of the rainy season in June, this tardy decision compromised the orderly movement of materials by sea and curtailed the prospects of building and employing the railway thereafter. In any future war, an effective wartime railway would depend upon operational and logistic circumstances.

It would benefit, too, from the engagement of engineers trained in railway management. In 1865 an investigatory committee had reported on the Royal Engineer Establishment, Chatham, where it found a steam engine constantly at work without officers being educated in its practical management. The school, renamed in 1869 as 'The School of Military Engineering', duly expanded its syllabus and offered railway courses not only for officers but also for NCOs.[73] By 1878, instruction in field railways and trench tramways had become the largest single item in the school's instructional handbook. The latter included material and diagrams on theoretical and mathematical principles of railway design, the tasks of surveying, preparing the ground, handling curves, the permanent way, plate laying, the use of tools, laying the line, points, crossings, stations, signalling, calculations about the carriage of troops and the restoration and destruction of railway bridges. The manual also defined the railway duties of Royal Engineers as '1st To lay a short line of railway between two places. 2nd To restore existing line which has been damaged or destroyed by an enemy. 3rd To destroy an existing line as much as possible with given men and time.'[74] Royal Engineers, 'taught thoroughly' in 'locomotive, driving and other railway work',[75] would soon have opportunities to demonstrate their expertise in at least two of these three duties.

Notes

1 B. Bond (ed.), *Victorian Military Campaigns* (London: Hutchinson, 1967), appendix 1, pp. 308–11.

2 General Viscount Wolseley, 'The Standing Army of Great Britain', *Harper's New Monthly Magazine*, European edition, vol. 80 (1890), pp. 331–47, at p. 346.

3 The classic account of these wars, and the challenges posed by nature, is Col. C. E. Callwell, *Small Wars: A Textbook for Imperial Soldiers* (London: HMSO, 1896, reprinted London: Greenwood Books, 1990).

4 Pratt, *Rise of Rail-Power*, pp. 206–10; M. Robbins, 'The Balaklava Railway', *Journal of Transport History*, vol. 1, no. 1 (1953), pp. 28–43; Westwood, *Railways at War*, pp. 7, 83; B. Cooke, *The Grand Crimean Central Railway*, 2nd revised edition (Knutsford: Cavalier House, 1997); C. Wolmar, *Engines of War: How Wars were Won and Lost on the Railways* (London: Atlantic, 2010), pp. 23–9; P. Marsh, *Beatty's Railway* (Oxford: New Cherwell Press, 2000).

5 'We want more horses & wagons warm clothing & forage for the horses or they will all die when the cold & wet come': W. G. Romaine to C. J. Selwyn, 22 October 1854, in Major C. Robins (ed.), *Romaine's Crimean War: The Letters and Journal of William Govett Romaine Deputy Judge-Advocate to the Army of the East 1854–6* (Stroud, Gloucestershire: Sutton for the Army Records Society, 2005), pp. 32–4, at p. 34.

6 TNA, WO 33/2B, Horse Guards, memorandum, 8 February 1855.

7 Romaine to Selwyn, 18 November 1854, and Romaine to Lord Mulgrave, 6 December 1854, in Robins (ed.), *Romaine's Crimean War*, pp. 35 and 45; see also A. Massie, *The National Army Museum Book of the Crimean War: The Untold Stories* (London: Sidgwick & Jackson, 2004), ch. 6.

8 W. H. Russell, *The War: From the Landing at Gallipoli to the Death of Lord Raglan*, 2 vols. (London: George Routledge & Co., 1855), vol. 1, p. 280; see also Spiers, *Army and Society*, pp. 97–103.

9 TNA, WO 6/78, fos. 1–10, H. Roberts to Sir C. Trevelyan, 5 December 1854, and fos. 178–81, 24 January 1855. For a detailed account of the expenditure incurred by the Balaclava Railway, see TNA, WO 28/245; PP, *Railway (Balaclava to Sebastopol)* (1854–55), XXXII, 547, pp. 1–15; and Marsh, *Beatty's Railway*, pp. 1–11.

10 Lord Raglan to S. Herbert, 24 December 1854, in Lord Stanmore, *Sidney Herbert: Lord Herbert of Lea. A Memoir*, 2 vols. (London: John Murray, 1906), vol. 1, p. 311.

11 Wrottesley, *Life and Correspondence of Burgoyne*, vol. 2, p. 189.

12 Wrottesley (ed.), *Military Opinions of Burgoyne*, p. 209; Lt.-Col. A. Sterling, *The Highland Brigade in the Crimea: Founded on Letters Written during the Years 1854, 1855, and 1856* (Minneapolis: Absinthe Press, 1995), pp. 107–15; C. Fitzherbert, *Henry Clifford V.C.: His Letters and Sketches from the Crimea* (London: Michael Joseph, 1956), p. 166.

13 Burgoyne to Colonel Matson, 11 February 1855, in Wrottesley, *Life and Correspondence of Burgoyne*, vol. 2, pp. 218–19; Marsh, *Beatty's Railway*, p. 127.

14 Robbins, 'Balaklava Railway', p. 31; Wolmar, *Engines of War*, p. 23; Cooke, *Grand Crimean Central Railway*, p. 145; Marsh, *Beatty's Railway*, pp. 126–7.

15 Marsh, *Beatty's Railway*, pp. 91 and 97.

16 J. Beatty to Romaine, 8 February 1855, in Robins (ed.), *Romaine's Crimean War*, pp. 276–7;'The Balaclava and Sebastopol Railway', *Illustrated London News*, vol. 26, no. 742 (12 May 1855), p. 450: a copy of Beatty's report is included in Cooke, *Grand Crimean Central Railway*, appendix 1, pp. 164–70, at pp. 165–6.

17 Marsh, *Beatty's Railway*, p. 126; Fitzherbert, *Henry Clifford*, pp. 171, 179–80; Maj. C. Robins (ed.), *The Murder of a Regiment: Winter Sketches from the Crimea 1854–1855 by an Officer of the 46th Foot (South Devonshire Regt) Annotated, and with Additional Material* (Bowdon: Withycut House, 1994), p. 25.

18 V. Bonham Carter (ed.), *Surgeon in the Crimea: The Experiences of George Lawson Recorded in Letters to his Family 1854–1855* (London: Constable, 1968), p. 160.

19 Russell, *The War*, vol. 1, p. 334; for a good selection of the contemporary images, see Cooke, *Grand Crimean Central Railway*, plates 4–22; 'Progress of the Balaclava

Railway to the Church of Kadikoi', *Illustrated London News*, vol. 26, no. 733 (24 March 1855), p. 272.

20 Romaine to C. J. Selwyn, 11 March 1855, in Robins (ed.), *Romaine's Crimean War*, p. 93.

21 TNA, WO 33/2A, Lord Panmure to the duke of Cambridge on the campaign of 1856, enclosing Col. E. Wetherall, report, 14 January 1856; Bonham Carter (ed.), *Surgeon in the Crimea*, p. 161; Capt. C. E. Luard, 'Field Railways, and their General Application in War', *Journal of the Royal United Service Institution*, vol. 17 (1873), pp. 693–724, at p. 694; Cooke, *Grand Crimean Central Railway*, p. 64.

22 Peto would claim that 39¼ miles of track were laid (A. Helps, *Life and Labours of Mr. Brassey* (London: Bell and Daldy, 1872), p. 217), but this may refer to the track sent out to the Crimea; see Cooke, *Grand Crimean Central Railway*, pp. 119–21, 135–6; Wolmar, *Engines of War*, pp. 25–6, and on the locomotives, University of Leeds, Liddle Collection, Mann MSS, G. O. Mann to his parents, 26 October, 1855, 28 December 1855 and 4 January 1856.

23 Russell, *The War*, vol. 2, pp. 218 and 359. On the operation of the railway, see TNA, WO 33/2A, Wetherall, report, 14 January 1856; Wolmar, *Engines of War*, p. 26, and Cooke, *Grand Crimean Central Railway*, p. 135. One of the worst accidents occurred on 6 April 1855; it killed a soldier of the 71st Foot and injured several others as well as Beatty himself, ultimately causing the latter's departure from the Crimea and his death at the age of thirty-six; see Marsh, *Beatty's Railway*, pp. 165–7, and Lord Raglan to Lord Panmure, 7 April 1855, in Sir G. Douglas, Bart., and Sir G. D. Ramsay (eds.), *The Panmure Papers: Being A Selection from the Correspondence of Fox Maule, Second Baron Panmure, afterwards Eleventh Earl of Dalhousie, K.T., G.C.B.*, 2 vols. (London: Hodder & Stoughton, 1908), vol. 1, p. 144.

24 Cooke, *Grand Crimean Central Railway*, ch. 9, especially pp. 146–8. On the railway's role in supporting the artillery, see Russell, *The War*, vol. 1, p. 402; Wolmar, *Engines of War*, pp. 26–7; O. Figes, *Crimea: The Last Crusade* (London: Allen Lane, 2010), p. 256; Massie, *National Army Museum Book*, p. 244.

25 TNA, WO 33/2B, Sir W. Codrington to Lord Panmure, 10 December 1855, enclosing Col. E. Wetherall, report, 5 December 1855, p. 10; Romaine to Selwyn, 21 May 1855, and Romaine to Lord Mulgrave, 11 September 1855, in Robins, *Romaine's Crimean War*, pp. 143 and 212–13; Wolmar, *Engines of War*, p. 27.

26 Sterling, *Highland Brigade*, p. 123; Bonham-Carter (ed.), *Surgeon in the Crimea*, p. 161.

27 None of their animals were under shelter by November 1855: TNA, WO 33/2A, Wetherall, report, 14 January 1856; Cooke, *Grand Crimean Central Railway*, pp. 129–30; see also Massie, *National Army Museum Book*, p. 244.

28 Duke of Newcastle, diary, 27 July 1855 in J. Martineau, *The Life of Henry Pelham Fifth Duke of Newcastle 1811–1864* (London: John Murray, 1908), p. 268; on the railway's performance, see Westwood, *Railways at War*, p. 7; Cooke, *Grand Crimean Central Railway*, pp. 126–35; Wolmar, *Engines of War*, p. 28.

29 Russell, *The War*, vol. 2, p. 429.

30 Quoted in Helps, *Life and Labours of Mr. Brassey*, p. 217.

31 Wolmar, *Engines of War*, pp. 28–9; Cooke, *Grand Crimean Central Railway*, pp. 150–1.

32 TNA, WO 33/2B, Codrington to Panmure; see also Sir J. Simpson to Panmure, 6 October 1855, and Codrington to Panmure, 1, 11 and 28 December 1855, in Douglas and Ramsay (eds.), *Panmure Papers*, vol. 1, p. 429, and vol. 2, pp. 3, 16 and 38.

33 Only one regiment of the Madras army refused to serve the British, and of the Bombay army, only elements of two infantry regiments and two artillery companies mutinied. Ultimately the British had the assistance of twenty Indian 'followers', official and unofficial, for every one British soldier: T. A. Heathcote, *The Military in British India: The Development of British Land Forces in South Asia 1600–1947* (Manchester: Manchester University Press, 1995), p. 106; David, *Indian Mutiny*, pp. 19, 232–3, 235–7.

34 P. Mason, *A Matter of Honour: An Account of the Indian Army, its Officers and Men* (London: Jonathan Cape, 1974), p. 281.

35 R. Llewellyn-Jones, *The Great Uprising in India, 1857–58: Untold Stories, Indian and British* (Woodbridge: The Boydell Press, 2007), p. 50; I. J. Kerr, *Building the Railways of the Raj 1850–1900* (Delhi: Oxford University Press, 1997), p. 35; M. Satow and R. Desmond, *Railways of the Raj* (London: Scolar Press, 1980), pp. 14–16; and Lord Dalhousie to the Court of Directors of the East India Company, 20 April 1853, in S. Settar and B. Misra (eds.), *Railway Construction in India: Select Documents*, 2 vols. (New Delhi: Northern Book Centre/Indian Council of Historical Research, 1999), vol. 2, pp. 23–57.

36 J. W. Kaye, *Lives of Indian Officers*, 3 vols. (London: W. H. Allen, 1867), vol. 3, pp. 225–6.

37 Ibid.; see also M. Edwardes, *Red Year: The Indian Rebellion of 1857* (London: Cardinal, 1975), p. 81, and C. Hibbert, *The Great Mutiny: India 1857* (London: Allen Lane, 1978), p. 200.

38 Llewellyn-Jones, *Great Uprising*, pp. 50–1; see also G. Huddleston, *History of the East Indian Railway* (Calcutta: Thacker, Spink and Co., 1906), p. 20.

39 Satow and Desmond, *Railways of the Raj*, p. 16; see also Huddleston, *History of the East Indian Railway*, pp. 19–20.

40 The Russians had moved a corps of 14,500 men, complete with all its horses and transport, over 200 miles by rail over two days in 1846, and then Austria transferred 75,000 men by railway from Hungary and Vienna to Bohemia in 1850, but during the eighty-six days of this war, the French railways transported 604,381 men and over 129,227 horses. Some 227,649 men and 36,357 horses went directly to the theatre of operations in Italy. M. van Creveld, *Supplying War: Logistics from Wallenstein to Patton* (Cambridge: Cambridge University Press, 1977), p. 82.

41 Maj. Miller, 'The Italian Campaign of 1859', *Journal of the Royal United Service Institution*, vol. 5 (1861), pp. 269–308, at p. 269.

42 Ibid. pp. 270–1.

43 This led to an armistice between France and Austria at Villafranca (11 July 1859), whereby Austria ceded most of Lombardy to France, whereupon France planned to cede these territories to Sardinia. This covert deal caused outrage in Sardinia (and provoked Cavour's resignation), but was overtaken by further Piedmontese gains within the central Italian states.

44 Miller, 'Italian Campaign', pp. 271, 276, 278–9.

45 Capt. H. W. Tyler, 'Railways Strategically Considered', *Journal of the Royal United Service Institution*, vol. 8 (1865), pp. 321–43, at pp. 325–32.

46 Luard, 'Field Railways', p. 694; see also Wolmar, *Engines of War*, pp. 30–1.

47 Tyler, 'Railways Strategically Considered', p. 332. This scale broadly encompasses the Confederacy (see a map transposing the Confederate states on central Europe in J. E. Clark, Jr, *Railroads in the Civil War* (Baton Rouge: Louisiana State University Press, 2001), p. 29), but the war was fought principally in the eastern and western theatres, with many of the principal battles fought in the east while Confederate losses in the west ultimately ensured their defeat: Wolmar, *Engines of War*, p. 42.

48 The British observers of the Civil War included official parties, mainly from the corps of Royal Engineers or Royal Artillery interested in the technical aspects of the war, and unofficial observers, including Wolseley. The latter commented principally upon the role of cavalry, the use of entrenchments, the quality of the troops and the unusual terrain, often doubting that the war had produced many lessons. Their interest in the Civil War was soon submerged by popular fascination with Prussian military success. J. Luvaas, *The Military Legacy of the Civil War: The European Inheritance* (Chicago: University Press of Kansas, 1959), pp. 8–19, 22–3, 34–7, 46–9,100–7, 116–17.

49 Tyler, 'Railways Strategically Considered', pp. 333–42.

50 There is a massive bibliography on the use of railways during the American Civil War but among the classical accounts are R. C. Black III, *The Railroads of the Confederacy* (Chapel Hill: University of North Carolina Press, 1952); J. F. Stover, *American Railroads*, 2nd edition (Chicago: University of Chicago Press, 1997); G. E. Turner, *Victory Rode the Rails* (New York: Bobbs-Merrill, 1953); T. Weber, *The Northern Railroads in the Civil War* (New York: King's Crown, 1952).

51 Clark, *Railroads in the Civil War*, p. 22, and on the Southern shortcomings, see pp. 6, 21, 24, 218–21, 223.
52 Luard, 'Field Railways', pp. 694–7, at p. 696.
53 RA, VIC/ADD MSS E/1/5575, Cambridge MSS, Sir R. Napier to the duke of Cambridge, 24 August 1867.
54 Maj. T. J. Holland and Capt. H. Hozier, *Record of the Expedition to Abyssinia, Compiled by Order of the Secretary of State for War*, 2 vols. (London: Her Majesty's Stationery Office, 1870), vol. 2, p. 339; see also Lt. Willans, 'The Abyssinian Railway', *Papers on Subjects Connected with the Duties of the Corps of Royal Engineers*, new series, vol. 18 (1870), pp. 163–76.
55 Apart from Lieutenant Willans, Lieutenant Pennefather, RE, served on the railway throughout the campaign, Captain Darrah, RE, supervised the railway works, and Lieutenant Baird, RE, assumed the duties of traffic manager from the beginning of March: Lt.-Col. Henry St Clair Wilkins, 'Abyssinian Expedition', report to Capt. T. S. Holland, *Papers on Subjects Connected with the Duties of the Corps of Royal Engineers*, new series, vol. 17 (1869), pp. 140–8, at p. 143.
56 Holland and Hozier, *Record of the Expedition to Abyssinia*, vol. 2, p. 339.
57 D. Bates, *The Abyssinian Difficulty: The Emperor Theodorus and the Magdala Campaign 1867–68* (Oxford: Oxford University Press, 1979), p. 130.
58 G. A. Henty, *The March to Magdala* (London, 1868, reprinted London: Robinson Books, 2002), pp. 52, 161. For sketches and engravings of spectacular scenes in the Abyssinian Highlands, British forces in action, Magdala burning, local peoples and elephant trains but not the railway, see the works of Roger Acton and Major Baigrie in the *Illustrated London News*, vols. 52 and 53 (4 January – 1 August 1868).
59 Henty, *March to Magdala*, pp. 61, 150–1, 156.
60 Wilkins, 'Abyssinian Expedition', p. 144.
61 Ibid.; PP, Capt. George Tryon (Qs. 2023–4), in *Report from the Select Committee on the Abyssinian War: Together with the Proceedings of the Committee, Minutes of Evidence, and Appendix*, no. 380 (1868–69), VI, and Willans, 'The Abyssinian Railway', pp. 165–6.
62 G. R. Elsmie (ed.), *Field-Marshal Sir Donald Stewart: An Account of his Life, Mainly in his Own Words* (London: John Murray, 1903), p. 158.
63 Wilkins, 'Abyssinian Expedition', p. 142; Willans, 'The Abyssinian Railway', p. 167.
64 Wilkins, 'Abyssinian Expedition', p. 142; for a useful account of the railway's difficulties, see D. G. Chandler, 'The Expedition to Abyssinia 1867–8' in Bond (ed.), *Victorian Military Campaigns*, pp. 107–59, at pp. 125 and 131.
65 Willans, 'The Abyssinian Railway', pp. 169, 171 and 173; Holland and Hozier, *Record of the Expedition to Abyssinia*, vol. 2, pp. 19, 108; H. M. Stanley, *Coomassie and Magdala: The Story of Two British Campaigns in Africa* (London: Sampson Low, Marston Low & Searle, 1874), p. 505.
66 Willans, 'The Abyssinian Railway', p. 176; Holland and Hozier, *Record of the Expedition to Abyssinia*, vol. 2, pp. 108 and 342.
67 PP, Tryon (Q. 2019), 'Select Committee on the Abyssinian War'.
68 RA, VIC/ADD MSS E/1/5592, Cambridge MSS, Napier to the duke of Cambridge, 13 October 1867.
69 RA, VIC/ADD MSS E/1/5840, Cambridge MSS, Napier to the duke of Cambridge, 28 May 1868.
70 Wilkins, 'Abyssinian Expedition', p. 144; Holland and Hozier, *Record of the Expedition to Abyssinia*, vol. 2, pp. 341–2.
71 Willans, 'The Abyssinian Railway', pp. 172–4.
72 Wilkins, 'Abyssinian Expedition', p. 144.
73 Maj.-Gen. W. Porter, *History of the Corps of Royal Engineers*, vol. 2 (Chatham: Institution of Royal Engineers, 1889), pp. 183–4, 187.
74 School of Military Engineering, Chatham, *Instruction in Military Engineering*, vol. 1, part V (London: HMSO, 1878), pp. 14–45.
75 Porter, *History of the Royal Engineers*, vol. 2, p. 183.

4

Operational railways

If the American Civil War demonstrated the military value of railways, Prussian usage – first in a sudden defeat of Austria (1866) and then in the spectacular and unexpected rout of France (1870–71) – confirmed expectations in Europe. The surrender at Sedan of the Châlons army under Marshal Patrice MacMahon, accompanied by Napoleon III, a mere six weeks after the declaration of war, constituted 'astounding intelligence', wrote the *Glasgow Herald*: 'the greatest military Power in the world' had suffered a 'sudden and complete collapse'.[1] British commentators, though aware that the strategic imperatives of Continental armies differed sharply from the imperial priorities of the British army,[2] recognized that the military had to take heed of these developments. At a minimum it had to construct organizational and technical capabilities in railway management. This would involve not merely reflection upon any lessons from the Franco-Prussian War but also continued experiments with the operational use of railways during expeditionary campaigns in Africa.

Commentary on the Franco-Prussian War

In a lecture before the RUSI on 12 May 1873, Captain C. E. Luard, RE, reviewed the general application of field railways in recent wars. 'Both sides', he argued, had begun the Franco-Prussian War 'with considerable experience of the working of railways, and poured their troops and supplies on to the frontier with all speed'.[3] This was certainly true, and the French probably had a better rail system with faster trains operating on a standardized network (unlike the German system, which was marred by political difficulties). The French could run more trains per day because they possessed a greater proportion of double-track lines, more rolling stock, and a larger capacity (for unloading) at their railway stations. Elements of their forces actually reached the frontier first in July 1870 but, as the French tried to save time by combining mobilization and concentration, massive confusion ensued. The Prussians and the armies of the southern German states had demonstrated their grasp of railway-borne deployments in 1866 when they deployed 197,000 men, 55,000 horses and 5,300 vehicles by rail within twenty-one days, a feat exceeded in 1870 when about 462,000 soldiers reached the French frontier within eighteen days.[4] Luard appreciated that the Germans had placed 'complete' forces on the frontier, 'and did it methodically . . . whereas the French sent everybody labeled *à Berlin*, and much confusion resulted when they arrived at the terminus of the line'. However, he then erred in a way that many others would do by claiming that 'the German organization appears to have been more perfect throughout than that of the French, and the results were consequently more successful'.[5]

In fact the Prussian-led forces had made numerous mistakes, albeit fewer than their enemies, and had not even learned from the organizational difficulties encountered by their rail-borne movements in 1866. In that war, and in 1870, the Prussians sent supply trains after the despatch of their combat forces, and, in both wars, these loaded supply trucks caused massive congestion because they were sent forward without consideration for the unloading facilities at the end of their journeys.[6] Spread over a broad front in 1870, the German armies had to live off the land as long as they manoeuvred in the mobile phase of the operations, but, as soon as they halted and embarked upon siege operations at Metz and later Paris, their supply problems became acute. What impressed Luard, and here his paper becomes a source for modern scholarship,[7] was the Prussian organization and management of their railway skills,[8] which enabled them to construct and maintain a field railway round Metz. Following a survey of the ground on 14 August, construction of the railway began two days later and was finished by the end of September. Stretching some 36 km, with two bridges over the Seille and Moselle and two viaducts, the railway was constructed at a rate of about 914 m per day. It required the daily efforts of some 4,000 men, composed of two field railway companies (450 men), four fortress pioneer companies (800 men) and about 3,000 unemployed coal miners. They used 250 vehicles and later eighty-four wagons from the pioneer trains of the 7th and 8th army corps.

When Metz surrendered on 27 October, the railway had had only twenty-six days of usage, and during the autumn rains it had required almost as many men to maintain the line as to build it. Although the bulk of the line ran over relatively flat country, the ruling gradient was an incline of 1 in 40, limiting capacity to a maximum of four vehicles per train. 'The railway', argued Luard, was not 'particularly well constructed', and on account of its sharp curves, 'traction became very dangerous and several accidents took place'. The fall of Metz, he claimed, had 'relieved the German engineers from their difficulties', but he drew the following lessons from the German experience:

> without a recognized military railway department, war cannot be carried on against first class powers at the present time with any prospect of success . . . [and] against all nations, whether civilized or uncivilized, the application of railways as a branch of military operations advances with rapidity the conclusion of such war, and consequently repays its own cost to an immeasurable extent.[9]

Wolseley also drew railway lessons from the Franco-Prussian War as well as the American Civil War in reissues of his *Soldier's Pocket Book*, a practical guide for soldiers on field service. Convinced that railroads would become 'the main lines of supply' in wars between 'civilized countries', and a valuable adjunct in other wars where conditions suited, he advocated laying down railways and using them wherever possible. Any movements over English rail, he argued, 'where rolling stock is practically unlimited' and double track available, meant that an army corps could move 600 sabres and bayonets per mile. The trains would probably be about half the length of their German counterparts owing to the length of British sidings and platforms, but they could carry men, ordnance, horses and stores. Wolseley prescribed, too, the procedures which should be followed in the march to the station, embarkation, the journey, disembarkation and the subsequent march to the bivouac or camping ground. He was particularly concerned that discipline must be enforced 'most strictly while marching through a town to the railway' lest there be any recurrence of the disorders that 'took place in 1870 at Paris, Metz, Amiens, and other great termini'. On the journey, a detachment of railway workmen 'with 20 or 30 rails should be with the leading train' in case of attacks on the line, and the carriages should contain skids or broughs for the rapid disembarkation of horses, guns and wagons between stations (as required in Canada in 1867 to respond to Fenian raiders). Finally, he insisted that any disembarkation had to be complete and speedy, with neither troops nor stores left near the point of disembarkation to minimize the chances of congestion or accidents: 'Any block at the place of arrival is more serious even than at the point of departure, for it must jam up all trains in the rear . . .'[10]

African campaigns, 1873–85

(a) Wars with the Asante, Zulu and Transvaal Boers

Wolseley found scant opportunity to employ these precepts in Africa, but he considered using a field railway in support of his expeditionary force in the Asante War of 1873–74. After heeding the advice of

Sir John Lintorn Simmons,[11] the lieutenant-governor of the Royal Military College, Sandhurst, Wolseley reckoned that the construction of a 48-km, narrow-gauge railway would 'shorten the duration of the operations and would render their execution more sure'.[12] At a cost of £45,000, two steam traction engines,[13] each weighing four tons, were made available with four tenders, together with sufficient rails for up to 13 km initially, and with another engine, four tenders, light trucks, ten trucks and up to 64 km of rail to be supplied subsequently if the line proved practicable. By 7 October, the initial material was assembled at Woolwich, and by 27 October, additional material was available for despatch. However, almost as soon as he arrived at Cape Coast Castle, Wolseley was briefed about the unevenness of the ground and the difficulties of laying a railway through the tropical rain forest within the limited time available. He duly informed Edward Cardwell, the secretary of state for war, that the country was 'unsuitable for railroad and traction engines' and 'that no further steps should be taken to provide a railroad'.[14] He later apprised the duke of Cambridge about the steepness of the gradients in many places and the lack of time, which precluded a railroad option.[15]

Neither of the next two African campaigns afforded any opportunity to construct and utilize a field railway. Hostilities with the Zulus were triggered by an ultimatum drafted by Britain's 'man on the spot', Sir Bartle Frere, the high commissioner for South Africa. When the terms of the ultimatum expired after thirty days, British forces under Lieutenant-General Lord Chelmsford crossed the border into Zululand on 11 January 1879, and part of his force encountered Zulus at the disastrous battle of Isandlwana (22 January 1879). Despite Zulu defeats at Khambula (29 March) and Gingindlovu (2 April), Chelmsford had to launch a second invasion with much larger forces in June 1879, and found himself under pressure to conclude the war quickly before Wolseley superseded him in overall command. On 4 July 1879 Chelmsford won a decisive victory at Ulundi and destroyed the kraal of the Zulu king, Cetshwayo.

After the British defeat of their enemies, the Zulu and later the Pedi in the Eastern Transvaal, the Transvaal Boers became increasingly restive about the British annexation of their country since 1877. On 16 December 1880 their protests erupted into armed confrontation, followed by an attack on a British convoy escorted by the 94th Regiment of Foot (later the 2nd Battalion, Connaught Rangers) at Bronkhorstpruit (20 December 1880). Some minor British reverses (Laing's Nek on 28 January and Schuinshoogte on 8 February) preceded the humiliating defeat on Majuba Hill (27 February 1881). The latter prompted Gladstone's government to sue for peace and ratify the agreement of

23 March with the Pretoria Convention (3 August 1881), which granted the Boers self-government under British suzerainty.

In neither war was there time to build a field railway, and in neither war were expeditionary forces sent from Britain. But reinforcements were sent to both conflicts, and they had to be moved to the front, posing an additional burden on the transport and supply arrangements, particularly in the Anglo-Zulu War. In this conflict, the costly and cumbersome supply and transport arrangements incurred ferocious criticism both at the time[16] and subsequently.[17] Chelmsford was even criticized for failing to use the railway lines out of Durban,[18] but Commissary-General Edward Strickland had persuaded the railway owners to allow him to use the few kilometres of unopened line, and the War Department had priority on the line out of Durban from January onwards.[19] The railway, though, was only partially built, and, contrary to some claims,[20] it never enabled troops to be moved by train up to Pietermaritzburg. In fact, the rails reached the Pietermaritzburg yard only on 21 October 1880, with a formal opening of the Durban and Pietermaritzburg railway on 1 December 1880. The journey over the 113-km stretch of railway took six and a quarter hours and almost certainly enabled the reinforcements for the Anglo-Transvaal War, namely the 92nd Gordon Highlanders, who landed in Durban on 30 January 1881, to reach Pietermaritzburg on the following day.[21]

(b) Egyptian intervention (1882)

Planning was much more methodical when Gladstone's cabinet resolved to intervene in Egypt in 1882. Alarmed by the nationalist revolt led by Arabi Pasha, the Egyptian minister of war, and the reports from the 'men on the spot', Sir Edward Malet, the British consul-general in Cairo, and Sir Auckland Colvin, who along with his French counterpart was responsible for Egypt's 'financial credit', the cabinet was appalled by reports of anti-European riots in Alexandria (11 June 1882), the so-called 'massacre of Christians' and the flight of many Europeans.[22] The cabinet began planning to intervene against a military trained by Europeans and possessing notable skills in engineering and gunnery. A naval bombardment of fortresses at Alexandria on 12 July confirmed that resistance was likely, and, in the absence of European allies, the government prepared to despatch a large expeditionary force – some 35,000 men, drawn from garrisons in Britain, India and the Mediterranean.[23]

From the outset of its planning the British military intended to employ Egyptian railways to support the intervention, both in the initial diversionary operations in the vicinity of Alexandria and later in the drive from Ismailia on the western bank of the Suez Canal towards Cairo (see Map 3). On 3 July Wolseley formally proposed that the expedition should be able to use the Egyptian railway from Ismailia to Zagazig (where trains and other rolling stock could be seized) by taking five locomotives, five brake vans, 100 wagons and material for the construction of at least 16 km of railway: 'This will enable us to cut down our transport very much,' he asserted.[24]

Three days later Sir Andrew Clarke, the inspector-general of fortifications, appointed Major William A. J. Wallace, RE, who was the manager of the Bengal Northern Railway and then fortuitously home on leave, to command a military railway corps. Composed of seven officers, two buglers and ninety-seven NCOs and men, the corps possessed only eight experienced engine drivers. All the other men lacked experience in railway work but were chosen for their versatility: the corps amounted to about a half of the number that Wallace had requested. Known as the 8th (Railway) Company, RE, they received ten days' training on the London, Chatham and Dover Railway and some practical platelaying instruction from the chief engineers of the South Eastern and the London and South Western Railway Companies.

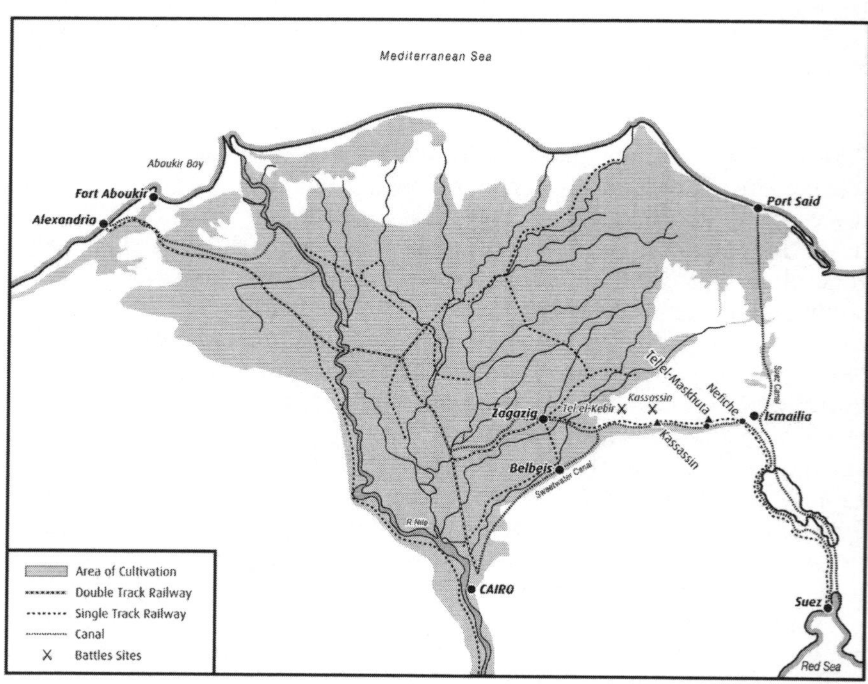

Map 3 Egyptian railways

These rushed preparations only confirmed that, at home, Royal Engineers fulfilled largely administrative duties in connection with British and Irish railways, with senior officers serving as railway inspectors, but in India, officers like Wallace had been able to use all their knowledge and skills.[25]

Wolseley, who would command the expedition, assured the queen that the 'fullest arrangements' had been made for regimental transport, and that the force, 'when it moves', would have 'the necessary railway plant to utilize the lines of railways along which it may have to move'.[26] The initial purchases lived up to this boast: the orders included four small tank engines, two first-class and two second-class carriages, six third-class carriages, forty cattle trucks, four brake wagons, two travelling cranes and two breakdown vans, and eight kilometres of permanent way complete with accessories and tools. While this plant was loaded on board the steamship *Lechmere*, Wallace travelled ahead to Alexandria, where he landed on 10 August, several weeks after the arrival of the advance units.

Skirmishes were already occurring along the railway line and its adjacent canal as British forces gave the impression that they were about to advance on Cairo from the north. Meanwhile Wallace secured another light engine from the Egyptian Railway Administration and bought another two tank locomotives for £800 apiece from Messrs Greenfield & Co. When the move from Alexandria got under way, he had the locomotive obtained from the Egyptian railway loaded onto the steamship *Osiris*, and nineteen wagons and some permanent way material loaded onto two large barges towed behind. One of those barges capsized off Aboukir – the 'fortune of war', as Wallace reflected – but four more locomotives, another nineteen wagons and more permanent way material were loaded onto a following vessel under the supervision of Captain D. A. Scott, the Commanding Royal Engineer, who had remained in Alexandria.[27]

Landing at Ismailia on the morning of the 21 August, two days ahead of his corps, would tax all the ingenuity of Wallace. He immediately took twelve sappers and about twenty sailors to repair two cuts in the railway line from Ismailia to Nefiche (about 4.8 km). At Nefiche station he found a train of twenty-two wagons at a time when he could not unload his own engine and materials because the *Recovery*, with its heavy landing gear, had not arrived at Ismailia. On the following day, Wallace raised a fatigue party of 400 men from the Guards and King's Royal Rifles to lay 914.4 m of track from the landing wharf at Ismailia to the railway. Raising fatigue parties, as Wallace admitted, 'was my greatest trouble throughout', especially because regiments were unloading their own kit at this time, and then subsequently, because the Pay Department refused to make additional payments for those who served on working parties.[28] Ordered by the Commanding Royal Engineer to take his locomotive and wagons on to Suez, Wallace did so and spent the next two days, with naval assistance, erecting the locomotive and wagons (with twelve of those forty-eight hours required to repair a burst boiler), mounting a 7-pounder gun on the leading truck, and fitting iron plating to protect the train from attack. The 8th (Railway) Company arrived at Ismailia late at night on 23 August, and the first service train towards the front ran on 28 August. Even then the train could not pass Tel-el-Maskhuta (about 13 km from Nefiche) because of a massive blockage on the line. It took another two days to remove the blockage, which was described by Wolseley as 'a great work' erected by the enemy,[29] and the first train, driven by Major Sidney Smith, RE, did not reach the base camp at Kassassin until 31 August.[30]

All these delays thwarted the first phase of the campaign. Wolseley had had to send his forces forward, trudging ankle-deep in sand across roadless terrain in fearful heat. As water carts struggled to keep up, soldiers in their desperation drank from the ill-named, and heavily polluted, Sweetwater Canal. The numbers of sick mounted rapidly, while units engaged the enemy in minor skirmishes and longer-range artillery duels. Lieutenant Charles Balfour (1st Battalion, Scots Guards) bemoaned the inability to bring forward sufficient medical stores and the fact that the Guards had 'no provision for the wounded'.[31] Medical support operating in single-lined tents, amid the intense heat of the day and severe cold at night, found itself beset by flies, periodically engulfed by sandstorms and bereft of horses whenever required to advance. Although it could, as Dr Alex S. Rose recalled, borrow horses to move its stores, the medical staff had to march until they fell exhausted. 'Unfortunately', he added, 'the transport service had broken down, and the result being that we were much hampered in all our movements, and sometimes were left quite helpless.'[32] Despite all his pre-war arrangements, Wolseley was forced to agree: 'As usual, transport has been our difficulty, and until we have an engine or two working on the line, I cannot move forward again.'[33] He admitted that the slow-moving steamship carrying the engines and plant from England had compounded his difficulties (the trains did not arrive until 5 September and were in service on the following day), and expected that 'the newspapers will howl at our delay' (as they did).[34]

Wolseley's luck held. The number of trains moving along the line on a daily basis steadily increased, with four or five engines hauling eight to fifteen wagons daily from 6 to 11 September, and then, when two larger engines arrived from Alexandria, trains of thirty-five loaded trucks were able to operate, delivering over 250 tons of supplies and baggage on a daily basis. Wolseley's forces resisted an Egyptian assault on their camp at Kassassin on 11 September 1882, and on the following night, they launched their famous night advance on the Egyptian entrenchments at Tel-el-Kebir. Faced with a massive pre-dawn attack on 13 September 1882, led by the Highland Brigade, the Egyptian forces broke within an hour, whereupon mounted infantry and the 4th Dragoon Guards pressed on to Cairo. Arriving at the Citadel at 5 p.m. on 14 September, these 120 men accepted the surrender of Arabi and his 8,000 soldiers. Thereafter, as Wolseley's forces secured the re-instatement of Khedive Tewfik and re-established British influence, Arabi was first sentenced to death and then exiled to Ceylon (now Sri Lanka).

The railway corps, sometimes assisted by a detachment of Madras Sappers, sustained the occupation. They opened a gap in the enemy lines at Tel-el-Kebir within two hours, and then pressed on to the station at Zagazig, where they captured another seven engines and eighty wagons. On 15 September they moved Wolseley on to Cairo, while telegraph communications were set up along the main line. Altogether from 28 August to 25 September, the British engineers despatched eighty-eight trains from Ismailia, carrying 1,308 truckloads or about 9,000 tons of supplies, stores and guns. They also forwarded to Cairo the large accumulations of stores that had been brought by rail and canal to Tel-el-Maskhuta and Kassassin, averaging 100 trucks or about 700 tons per day, and brought back the sick and wounded in return trains to Ismailia.[35] Childers was hugely impressed and proposed that the nucleus of a permanent railway corps should be established within the Royal Engineers. Sir Andrew Clarke endorsed the idea, as did Wolseley when adjutant-general in July 1883, so laying the groundwork for the formation of RE railway companies.[36]

(c) Sudan, 1884–85

Wolseley's enthusiasm for railways, however, waned when he planned the next major expedition in the Sudan, which followed the Mahdist insurrection and annihilation of a British-led Egyptian army at El Obeid (5 November 1883). As the rebels led by Mohammad Ahmed, the Mahdi or the 'Expected One', threatened further towns, including Khartoum, the Gladstone government, under popular pressure fanned by the *Pall Mall Gazette*, sent Major-General Charles 'Chinese' Gordon up the Nile to Khartoum. Gordon was a former governor-general of the Sudan, and he was instructed to 'consider and report' on the situation, including the possibility of evacuation.[37] While Gordon was *en route*, Egyptian forces under General Valentine Baker tried to relieve the beleaguered garrison of Tokar in the Eastern Sudan but suffered another crushing defeat at El Teb (4 February 1884), while a nearby garrison at Sinkat fell into rebel hands. Wolseley seized the initiative and in a lengthy memorandum (8 February) proposed that preparations for an intervention (which Gladstone and many of his cabinet desperately wished to avoid) should be made. He argued that any major intervention should be sent up the Nile and not across the desert from Suakin to Berber, and that additional British forces should be sent to relieve Tokar.[38]

In effect, Wolseley had triggered what became known as the 'battle of the routes"', with his preference for sending a flotilla of small boats up the Nile in a re-creation of his famous Red River expedition in Canada (1870). He was later supported by senior Red River veterans in the War Office, but challenged by intelligence officials, engineers, civil and military authorities in Cairo and senior naval authorities. The critics foresaw serious flaws in the Nile route, namely its great length, the vast number of animals and amount of forage required, the lack of sufficient boats and the problems presented by the cataracts. Accordingly, they advocated sending troops overland from Suakin to Berber and, in some cases, doing so by the construction of a railroad.[39]

Wolseley was then at the height of his prestige and influence. He had rightly discerned that 'Unless "something" is now done, and done at once . . . Gordon will soon find himself shut up in Khartoum',[40] and his support for intervention in Suakin temporarily changed the calculus of the debate. After Sir Gerald Graham, VC, led British forces to victories at El Teb (29 February 1884) and Tamai (13 March 1884), scattering the rebels in the Eastern Sudan, he massively boosted interest in the Suakin–Berber route. Wolseley, nonetheless, prepared lengthy memoranda (8 and 14 April among others) for Lord Hartington, the secretary of state for war, insisting that preparations must now to be made for a major expedition up the Nile. Hartington suspected that Wolseley had 'underrated the difficulties of the Nile route',[41] and support for the Suakin–Berber route came from Graham, Sir Charles Wilson and Lieutenant-Colonel John Ardagh (both in the intelligence department at Cairo), Sir Frederick Stephenson (commander of British forces, Egypt), Sir Evelyn Baring (agent and consul-general in Cairo), Lord John Hay (commander-in-chief of the Mediterranean Fleet), the Royal Engineers, and Sir Andrew Clarke and his protégé, but no relation, Captain George S. Clarke.[42]

The railway had already been mooted as a security option in Egypt not least by Gordon himself, and by an Egyptian commission in 1883.[43] The perceived virtues were not merely operational in the sense of rapidly moving troops and supplies into the interior but also political and commercial. Even the duke of Cambridge, though initially a supporter of the Nile route, maintained that 'a railroad from Suakin to Berber . . . is what is really wanted, and it would do more towards civilizing and settling that part of the country than anything else that could be devised'.[44] In other words, the railway was now being promoted as an instrument of strategy and political pacification.

On 19 May, Sir Andrew Clarke informed Lord Hartington that the railway from Suakin to Berber should be constructed 'at once'. He estimated the cost for 394 km of railway, including rolling stock, telegraph line and defensible stations every 16 km at £1,250,000. By plate laying at a rate of 3.2 km per day, he reckoned that 2,500 artificers, including 300 Europeans, could build it within about 130 days. Although slow to build, the railway would ultimately enable a much larger number of men to be moved forward at 'the last moment and their action rendered more rapid and certain'.[45] The first naval lord, Admiral Sir Astley Cooper Key, agreed: if the railroad could be completed by 1 November, he envisaged sending four or five steamers, in sections, up to Berber, whereupon they could be reassembled to assist an expedition to Khartoum. Even a partially completed railway 'from Suakin to the hills', he argued, would suffice; once it was completed, camels could be collected locally and a camel-borne expedition mounted. If complemented by a simultaneous expedition across the Korosko Desert, this would produce a 'tranquillizing effect throughout the country' and impede any northern advance by the Mahdi's forces.[46] However fanciful such speculation, Gladstone insisted that the railway was 'attended with the most formidable difficulties of a moral and political kind', but conceded that the 'Suakin and Berber route has utterly beaten Nile route for a larger expedition, & Railway has also fairly beaten camels'. He was still not prepared to endorse any relief expedition at this time, and wanted Gordon to withdraw voluntarily from Khartoum.[47]

Sir Andrew Clarke, who had political friends in the cabinet, continued his lobbying. He claimed that a railway would promote British 'indirect' interests in the Sudan, by preventing an aggressive Mahdist power from establishing itself in Khartoum, and by curbing the slave trade. On 6 June, he pressed the cabinet 'that no time should be lost'.[48] The cabinet, nevertheless, remained unconvinced about Gordon's predicament and about the need to send a relief force. War preparations proceeded incrementally. Just as on 10 May when Stephenson had received instructions to purchase camels and hold troops in readiness for a forward march in the autumn,[49] so on 14 June 1884 he received instructions to make preliminary preparations for the construction of the railway at Suakin. These included the building of a jetty, since a steamer with two locomotives, eight kilometres of rails and other rolling stock would be sent from Britain.[50] Unimpressed by this partial progress, Captain Clarke promoted the most ambitious of all rationales for the railway in an article in *The Times* of 18 July. Previous English expeditions, he argued, had proved 'costly', and lacked 'finality' because they seldom secured a 'permanent result'. Constructing a railway, though, could produce permanent political and military effects, with possible commercial advantages, thereby serving 'a constructive instead of a blankly destructive purpose'. Once a railway was made, he asserted, Berber would be within twelve hours of English ships, with easy access into the interior and no need to retain a permanent European garrison. 'The opening of the country to commerce', he insisted, 'is the one means by which tranquillity can be secured.'[51]

Wolseley kept lobbying Hartington, and another flurry of memoranda followed as Lord John Hay again deprecated the Nile route, while inside the War Office, Major-Generals Sir Redvers Buller and Sir John McNeill and Colonel William F. Butler, all Red River veterans, rallied in support of Wolseley.[52] But the die had been cast not by Wolseley's lobbying but by the fall of Berber to the Mahdists on 19 May, the news of which reached London three weeks later. When coupled with the dilatoriness of the government and the expense of the proposed railway, this development left the Nile route as the only practical proposition.[53] Following another threat to resign by Hartington on 31 July, the government requested a vote of credit on 5 August 1884. The ensuing expedition proved a disaster, with the small vanguard reaching Khartoum two days after it had fallen to the Mahdists and Gordon had been killed (26 January 1885). News of Gordon's death precipitated an outpouring of public grief, anguish and recriminations, producing a belated show of resolve from Gladstone's government. It ordered Wolseley to remain in the Sudan, and sent Graham back to Suakin with a much larger force to destroy Osman Digna's forces in the Eastern Sudan and to build the railway from Suakin to Berber (see Map 4).[54]

Meanwhile the 8th Railway Company had supported the Nile expedition, repairing the railway from Wadi Halfa to Sarras round the Second Cataract.[55] So poor was the condition of both the line and the locomotives that the railway made scant contribution to the advance of Wolseley's forces up the Nile, but the repair of 53 km of existing railway, and the construction of nearly 87 km of new line, assisted the evacuation in 1885.[56] Scenting his opportunity in the east of the Sudan, Sir Andrew Clarke undertook the initial arrangements for the Suakin–Berber railway. He had advocated a metre-gauge railway being laid by Indian labour under the supervision of the Indian Public Works

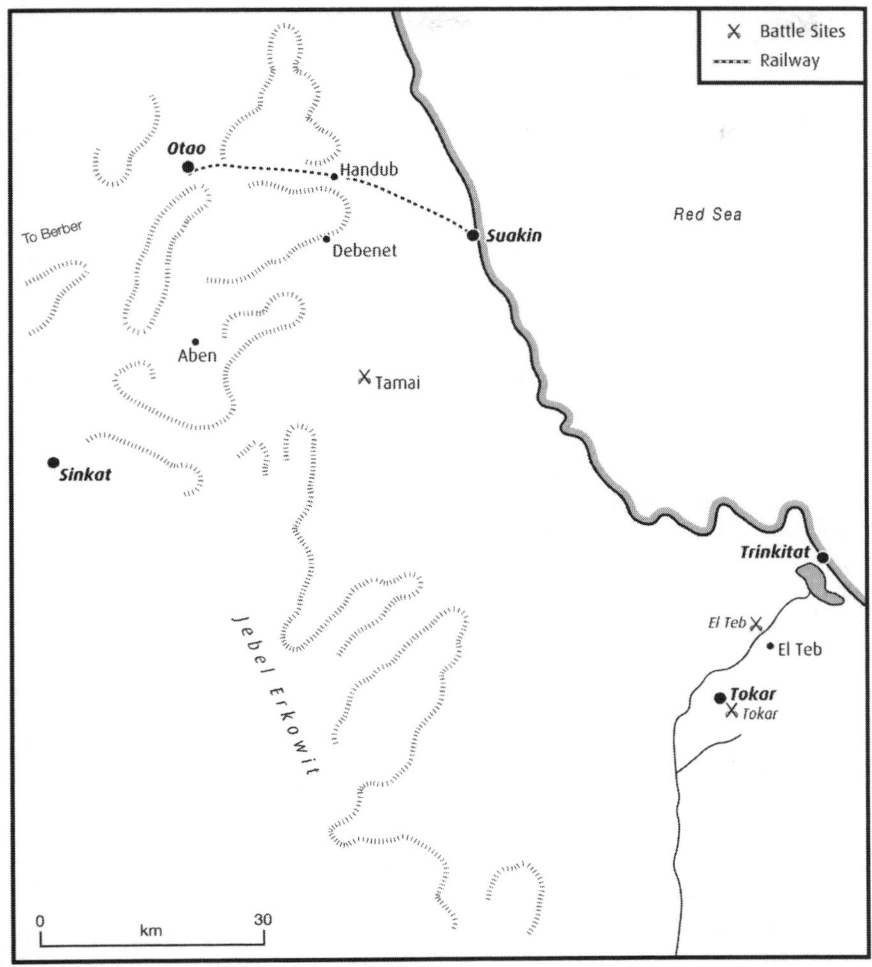

Map 4 East Sudan railway

Department but the Indian government objected to this diversion of resources from its own programme of railway construction, and so Sir Arthur Haliburton, who was director of supplies and transport at the War Office, put the work out to contract. Messrs Lucas and Aird promised to supply a railway of English standard gauge (4 feet $8^1/_2$ inches) and the accompanying supplies, and sent these materials in two vessels from London and Hull. Clarke also secured the services of the 10th Railway Company as well as forty men from Newcastle and Durham, and the 1st Lancashire Engineer Volunteers, whose trades could assist the railway work.[57]

However impressive on paper, these arrangements unravelled rapidly in the Eastern Sudan. Constant delays occurred at the railhead, where the ships were found to be improperly loaded – sometimes deficient in rails or sleepers – and rails, bolts and nuts were found to be of many different patterns. The civilian contractors, who were supposed to oversee all aspects of the work, and the 750 British workmen struggled in the local conditions. Despite free kit and pay at twelve shillings a day (or twelve times the daily pay of the ordinary soldier), the navvies barely completed the 'fishing' and spiking of the rails, while the contractors proved hopeless at organizing the project. The Royal Engineers, Madras Sappers and Indian labourers had to unload the ships and load the trains, survey the route, clear and level the land, lay out the line and manage the vital water supply. Construction began on 13 March but by 22 March only just over three kilometres had been laid, whereupon a hiatus followed as Graham sought to engage the enemy. Railway construction resumed only on 6 April, and by 13 April, it was proceeding at about 1.6 km a day. Three days later, the Scots Guards and 17th Railway Company, RE, with two artillery guns, seized Otao, while the troops remaining at Handub received reinforcements from Suakin.[58]

As the Mahdists continued to mount attacks at night, Graham sent reconnaissances in force out during the day, both in advance of the line and into the neighbouring valleys to clear them of hostile tribesmen. The 17th Railway Company, RE, had to arrange 'continuous defences' round camps and double the 'number of fortified outposts'.[59] By 30 April the line reached its ultimate terminus, Otao, nearly 31 km from the coast. On the same night an armoured train with two officers and fifteen Royal Engineers as guard came into service, patrolling the line between 21.00 hours and 04.00 hours and engaging an enemy that was pulling up rails, burning sleepers (see Figure 4) and cutting the telegraph line. However, 30 April was the day on which Wolseley insisted that the railway construction must be ended because the government had become more anxious about deteriorating relations with Russia. It took another fortnight before the government accepted that friendly tribes would not protect the line once the British withdrew, and so it was not until 18 May that Wolseley ordered all work on the line to cease and withdrawal to start immediately.[60]

In effect the line never came close to reaching Berber and was never used to supply the troops in the field. The failure produced bitter recriminations: Captain W. V. Constable, RE, who had had ten years

Figure 4 'Royal Engineers Clearing the Railway near Suakin of Burning Sleepers', *Cassell's History of the War in the Sudan*, 6 vols. (1885–86), vol. 5, p. 53

of railway experience in India, informed Wolseley that the railway line was

> The worst he had ever seen; not even the sidings on a contractors [*sic*] line are usually so bad. Instead of about 2000 sleepers per mile, there are not he says more than about 1200. The rails are laid level instead of with the usual cant inwards of about 1 in 20, and consequently all the wheels have very little bearing on the rails. The fish plates are not fitted to the rails: the engines cannot drag up more than seven carriages, & yet the gradients are not more than 1 in 100. Altogether a more pitiable attempt to construct a railway, or a more wretched railway when constructed it would be difficult to conceive.[61]

The Royal Engineers remained deeply vexed about the system of dual control and the employment of civilian contractors. Captain H. G. Kunhardt, RE, recalled that

> The line was laid out by a survey party of Royal Engineers; the formation was cleared and leveled by the Madras Sappers and Indian coolies under Royal Engineer officers; the P.W. material was unloaded from the ships, and the trains were loaded and unloaded by coolies, brought from India by, and working under the immediate orders of R.E. officers; the

> line was lifted, straightened, and boxed up by coolies and military labour under the immediate direction of R.E. officers. The carts and mules, the muleteers, transport horses and drivers, were all supplied by Government. The only work done by labour supplied by the civilian contractors was the running of the trains and the actual fishing and spiking down of the rails, for which the navvies were receiving from 12s to 15s a day pay, work which might have been done quite as well by the native plate-layers present on the spot for 1s a day.[62]

Quite apart from exposing the folly of dual control, and of making a mockery of the pre-war speculations about the cost, rate and feasibility of railway construction in a contested region, the Suakin–Berber railway was rightly dubbed 'a fiasco which cost the Government more than £865,000'.[63]

In short, over a period of a dozen years the British army had regularly considered railways as a means of facilitating their colonial operations. They had persuaded governments to invest in this option and to create the first specialist companies of railway (and telegraph) engineers. Any operational benefits, though, had depended entirely upon the nature and time-urgency of the overseas campaigns. The only real success had occurred in Egypt, where the Royal Engineers had exploited an existing rail network, and an adequate rolling stock, to support an expeditionary force over a relatively limited operational area. Where the local network was in a chronically dilapidated condition (as from Wadi Halfa), or the area of operation was still fiercely contested (as in the Eastern Sudan), any railway activity required ever-mounting resources and extensive amounts of time. The time and resources were unlikely to be made available unless the government had a direct interest in the political and strategic outcome. This was evident only on a fitful basis in the Sudan during 1884–85; the expeditions never received a sustained commitment and so the railway usage was never optimized.

Notes

1 Editorial, *Glasgow Herald*, 5 September 1870, p. 4.

2 'The Proper Basis for our Army', *Pall Mall Gazette*, 3 September 1870, p. 3; Col. J. S. Rothwell, 'The Conveyance of Troops by Railway', *United Service Magazine*, vol. 4, no. 757 (1891), pp. 213–21, at p. 217.

3 Luard, 'Field Railways', p. 701.

4 M. Howard, *The Franco-Prussian War*, 2nd edition (London: Routledge, 2002), p. 60; van Creveld, *Supplying War*, pp. 84, 86.

5 Luard, 'Field Railways', p. 701, and the comments of Major C. F. Powell on this paper, p. 718; see also *The War Correspondence of The Daily News 1870 edited, with Notes and Comments Forming a Continuous Narrative of the War between Germany and France*, 2 vols. (London: Macmillan, 1871), vol. 1, pp. 12 and 14.

6 Howard, *Franco-Prussian War*, pp. 69–70; van Creveld, *Supplying War*, pp. 84, 90, 107–8.
7 Van Creveld, *Supplying War*, pp. 91 and 259 n. 43.
8 For a detailed account of how the Prussian army developed the 'nuclei' of railway units from 1853 through to established railway battalions supporting each corps in the 1860s, and then the earliest use of railways in the Danish war of 1864, followed by the faltering efforts in Austria (1866), see van Creveld, *Supplying War*, pp. 78–80, 83–5.
9 Luard, 'Field Railways', pp. 703–4; on the washing away of the Moselle bridge on the day Metz capitulated, see Wolmar, *Engines of War*, p. 79.
10 Lt.-Gen. Sir G. J. Wolseley, *The Soldier's Pocket Book for Field Service*, 4th edition (London: Macmillan, 1882), pp. 441–65: see also Col. Sir Garnet Wolseley, *The Use of Railroads in War: A Lecture Delivered at Aldershot, on the 20th January 1873* (London: Edwin S. Boot, 1873).
11 For the expertise of Simmons on railways, see Chapter 2 n. 54.
12 TNA, CO 96/103, Sir G. Wolseley to the earl of Kimberley, 2 September 1873; see also I. F. W. Beckett (ed.), *Wolseley and Ashanti: The Asante War Journal and Correspondence of Major General Sir Garnet Wolseley 1873–1874* (Stroud, Gloucestershire: The History Press for the Army Records Society, 2009), pp. 26, 60–1.
13 Known as 'steam sappers', these engines were capable of operating on road or rail. The Rochester factory of Aveling and Porter first produced them in 1868. Beckett (ed.), *Wolseley and Ashanti*, p. 38.
14 TNA, WO 33/26, Wolseley to E. Cardwell, 13 October 1873.
15 RA, VIC/ADD MSS E/1/7207, Cambridge MSS, Wolseley to duke of Cambridge, 15–19 October 1873.
16 'War at the Cape: The Transport Difficulty', *[London] Daily News*, 22 May 1879, p. 5; 'The 21st Scots Fusiliers in Zululand', *Ayr Advertiser*, 26 June 1879, p. 5; J. C. Laband (ed.), *Lord Chelmsford's Zululand Campaign 1878–1879* (Stroud, Gloucestershire: Alan Sutton Publishing for the Army Records Society, 1994), pp. xv, xvii.
17 Laband (ed.), *Lord Chelmsford's Zululand Campaign*, p. xix; H. Bailes, 'Technology and Imperialism: A Case Study of the Victorian Army in Africa', *Victorian Studies*, vol. 24 (1980), pp. 83–104, at pp. 91–7.
18 Bailes, 'Technology and Imperialism', p. 92.
19 Lt.-Col. I. H. W. Bennett, *Eyewitness in Zululand: The Campaign Reminiscences of Colonel W. A. Dunne, CB South Africa, 1877–1881* (London: Greenhill Books, 1989), p. 50; B. Martin, 'The Opening of the Railway between Durban and Pietermaritzburg – 100 Years Ago', *Natalia*, vol. 10 (1980), pp. 35–40, at p. 36; 'Some Recollections of the Zulu War, 1879: Extracted from the Unpublished Reminiscences of the Late Lieut.-General Sir Edward Hutton, KCB, KCMG', *Army Quarterly*, vol. 26 (1928), pp. 65–80, at p. 66.
20 D. R. Morris, *The Washing of the Spears* (London: Sphere Books, 1968), p. 500.
21 Martin, 'The Opening of the Railway', pp. 35–40; Greenhill Gardyne, *Life of a Regiment*, vol. 2, p. 162.
22 H. C. G. Matthew, *Gladstone, 1875–1898* (Oxford: Clarendon Press, 1995), pp. 130–7; A. Schölch, 'The "Men on the Spot" and the English Occupation of Egypt in 1882', *Historical Journal*, vol. 19 (1976), pp. 773–85.
23 Duke of Cambridge to H. C. E. Childers, 13 July 1882, and Wolseley to Childers, 31 August 1882, in S. Childers, *The Life and Correspondence of the Right Hon. Hugh C. E. Childers, 1827–1896*, 2 vols. (London: John Murray, 1901), vol. 2, pp. 91–2 and 114–15; see also M. J. Williams, 'The Egyptian Campaign of 1882' in Bond (ed.), *Victorian Military Campaigns*, pp. 243–78.
24 Wolseley to the surveyor general of the ordnance, 3 July 1882, reproduced in Colonel J. F. Maurice, *Military History of the Campaign of 1882 in Egypt* (London: HMSO, 1887), pp. 4–5.
25 Porter, *History of the Royal Engineers*, vol. 2, p. 328; Maurice, *Military History*, pp. 7–8, 23; Maj. W. A. J. Wallace, 'Report of the Railway Operations in Egypt during

August and September 1882', *Professional Papers of the Corps of Royal Engineers*, vol. 9 (1883), pp. 79–98; PP, Wallace (Qs. 4629, 4633, 4641–2), in *Report from the Select Committee on Commissariat and Transport Services (Egyptian Campaign) together with the Proceedings of the Committee, Minutes of Evidence and Appendix*, C. 285 (1884), X, pp. 210–11.

26 Wolseley to Queen Victoria, 10 July 1882, in Childers, *Life and Correspondence*, vol. 2, pp. 94–5.

27 Wallace originally identified the *Canadian* as the slow-moving steamship but he corrected this in subsequent testimony: Wallace, 'Report of the Railway Operations', pp. 79–80; PP, Wallace (Qs. 4649, 4657–60, 4785), in *Report on Commissariat and Transport Services*, pp. 211 and 218.

28 PP, Wallace (Qs. 4684–6, 4698–9), in *Report on Commissariat and Transport Services*, pp. 212–13.

29 Wolseley to Childers, 31 August 1882, in Childers, *Life and Correspondence*, vol. 2, p. 114.

30 Quoting from the diary of Major S. Smith: Porter, *History of the Royal Engineers*, vol. 2, p. 160; Wallace, 'Report of the Railway Operations', pp. 81–2.

31 S. G. P. Ward (ed.), 'The Scots Guards in Egypt, 1882: The Letters of Lieutenant C. B. Balfour', *Journal of the Society for Army Historical Research*, vol. 51 (1973), pp. 80–104, at p. 91.

32 'Letter from a Townsman', *Brechin Advertiser*, 10 October 1882, p. 3.

33 Wolseley to Childers, 26 August 1882, in Childers, *Life and Correspondence*, vol. 2, p. 109.

34 Wolseley to Childers, 31 August 1882, in Childers, *Life and Correspondence*, vol. 2, pp. 114–15; see also Wallace, 'Report of the Railway Operations', p. 83, and *The Times*, 29 August 1882, p. 7.

35 Wallace, 'Report of the Railway Operations', pp. 84–5, 95.

36 Col. R. H. Vetch, *Life of Lieut.-General the Hon. Sir Andrew Clarke, G.C.M.G., C.B., CLE* (London: John Murray, 1905), pp. 239–40.

37 Earl of Cromer, *Modern Egypt*, 2 vols. (London: Macmillan, 1908), vol. 1, p. 443.

38 Wolseley memoranda, 8 February and 13 April 1884, in B. Holland, *The Life of Spencer Compton Eighth Duke of Devonshire*, 2 vols. (London: Longmans Green & Co., 1911), vol. 1, pp. 425–9 and 441–5.

39 There is an extensive literature on this controversy: Lt.-Col. E. W. C. Sandes, *The Royal Engineers in Egypt and the Sudan* (Chatham: Institution of Royal Engineers, 1937), pp. 88–91; A. Preston (ed.), *In Relief of Gordon: Lord Wolseley's Campaign Journal of the Khartoum Relief Expedition 1884–1885* (London: Hutchinson, 1967), pp. xxviii–xxxii; J. Symons, *England's Pride: The Story of the Gordon Relief Expedition* (London: Hamish Hamilton, 1965), pp. 65–72; Holland, *Life of Eighth Duke of Devonshire*, vol. 1, pp. 459–62, 464–72; H. Kochanski, *Sir Garnet Wolseley: Victorian Hero* (London: Hambledon Press, 1999), pp. 156–8; F. Nicoll, *Gladstone, Gordon and the Sudan Wars: The Battle over Imperial Intervention in the Victorian Age* (Barnsley: Pen & Sword, 2013), pp. 94–104.

40 Wolseley to Hartington, 8 February 1884, in Holland, *Life of Eighth Duke of Devonshire*, vol. 1, p. 426.

41 Lord Hartington to Sir F. Stephenson, 25 April 1884, in Sir F. C. A. Stephenson, *At Home and on the Battlefield: Letters from the Crimea, China and Egypt, 1854–1888* (London: John Murray, 1915), p. 39, and Holland, *Life of Eighth Duke of Devonshire*, vol. 1, pp. 450–1; see also TNA, WO 147/40, Wolseley memoranda, 8 and 14 April 1884.

42 TNA, CAB 37/12, Vice-Admiral Lord John Hay to Lord Hartington, 13 May 1884; WO 147/40, Lt.-Col. J. Ardagh, 'Suakin–Berber Railway', 5 May 1884; Col. R. H. Vetch, *Life, Letters, and Diaries of Lieut.-General Sir Gerald Graham, V.C., C.B., R.E.: With Portraits, Plans, and his Principal Despatches* (Edinburgh: Blackwood, 1901), pp. 274–5; RA, VIC/ADD MSS E/1/10738, Cambridge MSS, Stephenson to duke of Cambridge, 5 May 1884; Vetch, *Clarke*, pp. 265–6; Col. Lord Sydenham of Combe, *My Working Life* (London: John Murray, 1927), pp. 48–9.

43 BL, Add. MSS 34474 (vol. 1), fo. 54, Gordon MSS; Nicoll, *Gladstone, Gordon*, pp. 96–7; Augustus B. Wylde, *'83 to '87 in the Soudan: With an Account of Sir William Hewett's Mission to King John of Abyssinia*, 2 vols. (London: Remington, 1888), vol. 2, pp. 125, 130–5.
44 RA, VIC/ADD MSS E/1/10741, Cambridge MSS, duke of Cambridge to Stephenson, 9 May 1884; see also Sydenham of Combe, *My Working Life*, p. 48.
45 TNA, CAB 37/12, Sir A. Clarke to Lord Hartington, 19 May 1884.
46 Ibid. Sir A. Cooper Key to Lord Northbrook, 26 May 1884; see also Vice-Admiral P. H. Colomb, *Memoirs of Admiral the Right Honble. Sir Astley Cooper Key* (London: Methuen, 1898), pp. 467–9.
47 BL, Add. MSS 43875, fos. 169 and 171–2, Dilke MSS, W. E. Gladstone to C. W. Dilke, 28 and 30 May 1884.
48 Vetch, *Clarke*, pp. 265–7.
49 Stephenson, *Home and Battlefield*, p. 283.
50 TNA, WO 147/40, Lord Hartington to Stephenson, 14 June 1884.
51 'The Suakin–Berber Railway', *The Times*, 18 July 1884, p. 4; for the authorship of the article, see Sydenham of Combe, *My Working Life*, p. 48.
52 TNA, WO 147/40, Wolseley to Lord Hartington, 18 June, 19, 24 and 29 July 1884, enclosing memorandum by Major-Generals J. C. McNeill and Redvers Buller and Colonel W. F. Butler, 29 July 1884.
53 Kochanski, *Wolseley*, p. 157; Preston (ed.), *In Relief of Gordon*, p. xxxi; Symons, *England's Pride*, pp. 71–2.
54 TNA, WO 147/40, Lord Hartington to Wolseley, 6 February 1885; see also PP, *Egypt No. 2 (1885): Correspondence Respecting British Military Operations in the Soudan*, C. 4280 (1885), LXXXIX, pp. 7–8, No. 17, Lord Hartington to Wolseley, 9 February 1885; and Graham's instructions of 20 February 1885 in Vetch, *Graham*, p. 291.
55 For a description of how the platelaying was undertaken, see Appendix 1.
56 Sandes, *Royal Engineers in Egypt and Sudan*, p. 103; Lt. M. Nathan, 'The Sudan Military Railway', *Professional Papers of the Corps of Royal Engineers*, vol. 11 (1885), pp. 35–49.
57 Vetch, *Clarke*, pp. 269–71; Sandes, *Royal Engineers in Egypt and Sudan*, p. 69; J. B. Atlay, *Lord Haliburton: A Memoir of his Public Service* (Toronto: William Briggs, 1909), pp. 55–61.
58 B. Robson, *Fuzzy-Wuzzy: The Campaigns in the Eastern Sudan 1884–85* (Tunbridge Wells: Spellmount, 1993), pp. 155–7, 161, 168–9.
59 TNA, WO 32/6134, Report on Royal Engineers at Suakin, 15 April 1885, enclosing 'Diary of Events at Suakin, 17th Company Royal Engineers, 1st to 31st March 1885' by Lt.-Col. Elliot-Wood. On the problems that bedevilled the enterprise, see R. Hill, 'The Suakin–Berber Railway, 1885', *Sudan Notes and Records*, vol. 20, no. 1 (1937), pp. 107–24, and Robson, *Fuzzy-Wuzzy*, ch. 11.
60 TNA, WO 32 147/40, Lord Hartington to Wolseley, 13, 15 and 20 April 1885; see also Lord Hartington to Wolseley, 17 April 1885, in Holland, *Life of Eighth Duke of Devonshire*, vol. 2, pp. 33–6, at p. 34; see also Stephenson to his sister, 13 April 1885, in Stephenson, *Home and Battlefield*, p. 336, and Vetch, *Graham*, p. 302.
61 Wolseley diary, 15 May 1885 (with one correction, 'not' for 'nor' in the penultimate sentence), in Preston (ed.), *In Relief of Gordon*, pp. 211–12.
62 Capt. H. G. Kunhardt, 'Notes on the Suakin–Berber Railway', *Royal Engineers [RE] Journal*, vol. 15, no. 176 (1885), pp. 155–6, at p. 156.
63 Sandes, *Royal Engineers in Egypt and Sudan*, pp. 75–7, at p. 75; PP, *Suakim (Cost of Military Expedition)*, C. 360 (1884–85), XLVI, p. 2; Vetch, *Graham*, p. 303.

5

Strategic railways in India

The abrupt termination of military operations and railway building in the Eastern Sudan followed the rapid deterioration of Anglo-Russian relations after the Penjdeh Incident (23 March 1885), in which the Russians killed some 600 Afghans.[1] Arguably the ensuing rift was the closest that Britain and Russia came to war during the late nineteenth century as attention refocused upon the north-west frontier and the primacy of India's security among Britain's imperial concerns. Whether the Russian government, as distinct from some of its army officers, ever contemplated an attack on India provoked debate among both contemporaries and subsequent historians.[2] Doubtless, too, the perceived 'Russian threat' served disparate purposes; the Indian political authorities sometimes fuelled Russophobic fears to enhance the army's profile in imperial strategy and counter the navalist influence at home,[3] while senior army officers exploited India's defence as a means of securing loyalties, promotions and peerages.[4] Although diplomacy prevailed after the Penjdeh Incident, with Russia backing down and agreeing to a delimitation of the Russo-Afghan frontier, which was eventually resolved in 1887, the Indian authorities had to take the possibility of a Russian threat seriously, especially Russia's expansion in central Asia and the extension of its central Asian railway. Britain had twice intervened in Afghanistan – in 1839–42 and again in 1878–80 – in attempts to reduce the Russian influence in Kabul, and several British officers, including Sir Frederick Roberts, VC, who became commander-in-chief in India (1885–92), assumed that war with Russia was inevitable.[5] Irrespective of whether such a war might involve an offensive (as contingency planning assumed in the 1870s and early 1880s) or a more defensive approach (awaiting large reinforcements from Britain, which were never likely), British planning required a sustained improvement in the local road and rail communications. The extension of the railway up to Quetta, and thence to the Afghan border, became a strategic prerequisite.

Railway building had proceeded apace in India after the pioneering years of the 1850s and 1860s. British engineers had to overcome cultural differences between Indian workers and themselves, and between various groups of workers. They encountered diseases such as dysentery, typhus, cholera and malaria that decimated their workforces, the tardy receipt of construction equipment and supplies from Britain, challenging relationships with Indian officialdom, and the need to choose plant, materials and designs that would suit Indian conditions. Gradually, as Ian Kerr argues, trial and error, buttressed by guaranteed funding from the Indian government, evolved into more routinized processes 'based on substantial proven experience'.[6] Engineers found more assured supplies of labour; developed bridge-building techniques that stopped bridges from collapsing; addressed some of the causes of epidemic mortality (apart from malaria); and 'learned to live with government supervision', as the government began to build and operate railways on its own account (and assume majority ownership of the private railway companies after their first twenty-five years of operation). By 1880, India had 14,980 km of railways, and, by 1900, it had the fourth largest railway mileage in the world, some 40,396 km.[7]

Internal security

While political, social and potentially commercial impulses were to the fore in the bulk of Indian railway construction, military motives persisted even if they were downplayed somewhat naively by Horace Bell, a construction engineer with the East Indian Railway. 'The importance of the railway system in India for military purposes', he wrote,

> was naturally recognized at the outset, and great stress was laid on this in Lord Dalhousie's minute of 1853. But until the outbreak of the Afghan War, at the end of 1878, no comprehensive views had been taken of the interconnection of our frontier communications, nor any programme laid down for railway construction for purely or mainly military objects.[8]

Railways, however, could serve a multitude of purposes, especially if care was taken in the layout of the trunk-line construction. Following the Mutiny, which had exacerbated concerns about the security of the Raj, the rapid movement of troops by rail would be a crucial means of allaying such fears in the future. Exclusive access to railways enabled the imperial forces to concentrate units and arsenals near the main rail lines, and then to exploit a freedom of movement denied any rebellious parties. Accordingly, the great cities were linked with military cantonments, escorts were provided for railway construction gangs,[9] and provision was made to bolster the security of railway lines, bridges, tunnels and stations. While such measures were imperative on politically sensitive routes, such as the Amritsar–Delhi railway, and on key facilities like the Delhi Bridge,[10] the Government of India reckoned that precautions should be undertaken in regions less affected by the Mutiny. In Madras, for example, where stations had been built without protection in mind, the local authorities objected to the cost of adding new defences, doubted the practicality of defending stations and rolling stock when rails could be broken up, and feared the consequences of building defences that would evince 'a distrust of the Native population'.[11] The government, nonetheless, insisted that local authorities should not be deterred from 'taking a step which prudence commends as of considerable political and military importance'. It also described how relatively cheap defensive works could be added to stations so that they could be defended by railway employees until relief arrived, and any damaged lines were repaired.[12]

Physically, these precautions involved the design of many railway stations, bridges and tunnel entrances as fortified, defensible positions. The designs ranged from fairly primitive forms to structures that resembled medieval castles. In the north-west regions even small or third-class stations had minimal protective features, with the angles of the passenger building being loop-holed and a commanding tower built in connection with the well from which the water tower was supplied. Elsewhere some of the stations had strong enclosing walls with iron gates to close the openings, round corner towers, firing slits from which defenders could command the outer walls, and exteriors cleared of cover. Stations were seen as potentially places of refuge from which the occupants could be rescued by train-borne troops. Lahore station at the junction of the Lahore–Amritsar line was the most impressive of these edifices. Opened in 1864, it was not merely a formidable building but was also a psychologically dominant one – a symbol of Britain's power in India.[13]

In addition Europeans, Anglo-Indians (then known as Eurasians) and time-expired British soldiers, who had taken their discharge in India and often married, formed volunteer units to defend railway bridges, stations and parts of the rail tracks during disturbances. Motivated by memories of the Mutiny and the reports of the volunteer movement in the United Kingdom, these railway units were part of a much broader volunteer movement, encompassing rifle, artillery, light horse, engineer and signal corps. While many of these units were quite small and were duly absorbed or amalgamated into larger formations, the 'railway towns' spawned several larger formations. The East Indian Railway Volunteer Rifle Corps, raised in 1869, was the first railway unit, and many of the railway units were particularly strong because they recruited from 'ready-made communities and could recruit the entire length of their lines'.[14] By 1893, of the 26,746 Europeans and

Eurasians enrolled as volunteers, some 10,000 were railwaymen,[15] and many of the railway formations had absorbed smaller rifle corps or had been amalgamated into larger railway units (see Appendix 2). Despite their numerical strength, the railway corps were neither mobilized on a large scale nor utilized in any major emergencies before the end of the century. Even if a national emergency had occurred, as John Gaylor rightly observes, 'train-crews were too important as train-crews to allow them to act as soldiers'. In fact, para-military passions would oscillate over the years, and when the perception of an internal threat declined towards the end of the century, the railway units became 'little more than local clubs with military associations'.[16]

Nevertheless the Indian authorities monitored the potential usage of the railways for military transport, assessing rolling stock that could be converted to carry horses or the sick and wounded.[17] The Royal Engineers also tested the loading rates and embarkation techniques for military units. In 1875, a committee chaired by Lieutenant-General Sir Charles Reid reported on the embarkation rates of a squadron of the 10th Hussars, fully equipped for field service with 123 horses, camp equipage, ammunition and baggage. The force could be loaded in thirty-nine minutes, disembarked in twenty-eight minutes, and re-embarked in thirty-two minutes. A battery of horse artillery, with four mortars, seven wagons and 134 bullocks, could be loaded in forty minutes; the three guns could be disembarked and made ready for action beside the line in eight, thirteen and twenty-four minutes respectively before reloading was completed in twelve and a half minutes, 'showing thereby what could be done with a little practice'.[18] Railways in effect became a 'force multiplier' in India: as the Eden Commission reported in 1879, railways were 'a real addition to the military strength of the country', removing any need to scatter troops in 'small isolated detachments' across the country. 'So long as we hold our main trunk lines of railway communication', the commission asserted, 'any successful insurrection, or any real display of military power by the Native States, is out of the question.'[19] By 1889, Sir Frederick (later Lord) Roberts claimed that the army now needed 'fewer men . . . to quell disturbances. Then the railways which now intersect the country enable troops to be assembled with greater rapidity wherever their services may be required, the result being that there are definitely fewer disturbances than in former days.'[20]

Frontier railways

In the wake of the Mutiny, internal security remained the prime concern of the political and military authorities in India. Those sounding the tocsin over the 'Russian threat' found scant support from governments in the United Kingdom,[21] and the prevailing frontier policy of the 1860s and up to 1874 remained one of 'masterly inactivity'. It was closely identified with the preferences of some Liberal politicians, and was reaffirmed by the new Liberal administration of 1880–85 when it withdrew British forces from Kandahar in April 1881 (but retained them in Pishin and Sibi, with Quetta as the capital of the frontier province). This policy chimed with the preferences of Sir John Lawrence, who had been viceroy between 1864 and 1869, and other members of the so-called Lawrence school, notably Thomas Baring, first earl of Northbrook, another former viceroy (1872–76) and first lord of the admiralty (1880–85), and Lord Derby, secretary of state for India (1858–59) and foreign secretary (1870–74).[22]

A 'forward' imperial strategy, advocated by the Conservative government of Benjamin Disraeli, later the earl of Beaconsfield, found active promotion from Lord Lytton, the viceroy and governor-general of India (1876–80). Alarmed by reports of a Russian mission expanding its influence in Kabul, Lytton gained the support of a divided British cabinet for submitting an ultimatum to Sher Ali, the amir of Afghanistan, on 2 November 1878, thereby triggering the Second Anglo-Afghan War.[23] Lytton did not wish to occupy Afghanistan, and settled for a British mission in Kabul,[24] with control of Afghanistan's external relations at the Treaty of Gandamak (26 May 1879), but his aides and the Indian military authorities devised all manner of contingency policies, often involving offensives through Afghanistan, to check any Russian advance. Colonel George Pomeroy Colley, who was Lytton's military and later private secretary, envisaged a route through Peshawar and Balkh to Tashkent. Sir Frederick Haines, then commander-in-chief in India, and Roberts, his quartermaster-general, championed the route through Kandahar to Herat and Merv. In a series of twenty papers, drafted from 1877 to 1893, Roberts warned of an impending war with Russia and of India's vulnerability, and in many of these papers he promoted a forward defence of India along a 'scientific frontier' in Afghanistan, with a railway to Kandahar as a prime requirement.[25]

When Persia signed the Herat Convention (November 1879), agreeing to protect central Asia from Russian encroachment, British negotiators envisaged an extension of a railway from Kandahar to Herat and the improvement of the telegraphic communications within Persia. This was effectively an Anglo-Persian alliance but the concept was rejected by Disraeli's cabinet for fear of over-extending Britain's responsibilities.[26] In Afghanistan the new amir, Abdur Rahman Khan, wanted to recover Kandahar and restore his country's territorial integrity, an objective in keeping with the priorities of the incoming Liberal administration in London.[27] Following Roberts's famous march from Kabul to Kandahar, and his defeat of Afghan rebels at the battle of Kandahar (1 September 1880), Abdur Rahman secured his territorial aims by crushing the remaining rebels at Herat.

The war, nonetheless, had triggered the first development of the frontier railway. When the two divisions under Sir Donald Stewart marched northwards towards Kandahar in the winter of 1878–79, relying on camels and suffering heavy losses through sickness, a branch line was constructed from Ruk to Jacobabad and on to Sibi, near the Bolan Pass, which snaked northwards through the mountains to Quetta. Under Lieutenant-Colonel James Gavin Lindsay, RE, the engineer-in-chief, construction of the branch line began on 6 October 1879. The materials were gathered from all across India, and the 214-km railway across the desert from Sind to the Afghan mountains was completed at the unprecedented rate of a mile a day. It reached Sibi, which was simply a squalid collection of huts, on 14 January 1880,[28] and the line was opened on 27 January 1880. Preliminary work was then undertaken on what was known as the 'Kandahar State Railway' (KSR). This included a survey of the routes to Quetta, which gave preference, on account of its gradient, to the longer Harnai route over the Bolan route. Thereafter labourers pierced a tunnel and built quarters for the engineers as well as an earthwork,[29] but their work was halted by news of the disastrous defeat of British forces at Maiwand (27 July 1880). As the construction party and its military escort withdrew, a convoy was left exposed in the Kuchali defile (6 August 1880). Marri tribesmen fell upon it, killing forty-two men and seizing substantial amounts of money and treasure. Nevertheless, Gladstone's government confirmed the change of policy, abandoning Kandahar in April 1881.[30]

Central Asia, though, remained in turmoil as Russia extended its railway eastwards from the Caspian Sea and accelerated its advance towards Herat, often described as the 'gateway to India'.[31] Unwilling to ignore these movements, the Liberal government approved the resumption of road construction in 1883,[32] namely the Harnai Road Improvement Scheme, purging any reference to Kandahar. The project, though earmarked for military engineers, was placed under the Public Works Department with a ban on any use of rails or rolling stock. This retarded the rate of work because materials could not be transported on temporary railway lines, and increased the cost by relying upon expensive camel transport.[33] Fortunately the engineer-in-chief was a highly experienced man of great resource, energy and adaptability. Colonel (later Major-General Sir) James (Buster) Browne, Royal (Bengal) Engineers, was a Scot who had spent most of his active service in India, being wounded in the Mutiny and in the Umbeyla campaign (1863–64), where he was thrice mentioned in despatches. Having seen more active service in Afghanistan and Egypt, he became chief engineer in the Punjab, and when on furloughs from India in the early 1870s, he visited Holland and Belgium to learn about marine, bridge and railway engineering, and travelled to American cities to study bridge designs. He also taught as a professor of mathematics and earned a medal from the Institution of Civil Engineers. Blessed with remarkable linguistic skills, he gradually acquired knowledge of many of the languages and dialects on the frontier. In constructing the Sind–Pishin railway, he assembled a formidable team of engineers, known as his 'young bears'; they worked prodigiously on the line and in some cases paid the ultimate sacrifice for doing so. Browne, nonetheless, remained the driving force – an inspirational personality and a seemingly ubiquitous presence, 'controlling, advising, cheering the workers, comforting the sick'.[34]

Under his command were two (later three) battalions of pioneers and five companies of the Bengal Sappers and Miners, officered for the most part by young RE lieutenants. They were assisted by several engineers with railway experience, and by thousands of Indian labourers, described as 'the riff-raff and sweepings of other departments; for respectable natives would not go to a place with such an evil reputation, and good reliable men were not likely to be spared by their own masters'.[35] They faced an immense task. The route (see Map 5) rose from about 91 m above sea level at Sibi to the summit of the mountain pass, 1,981 m high, in a mere 193 km, and the entire route encompassed 360.5 km. The engineering would have to overcome daunting obstacles: the 22.5-km Nari gorge, through which the railway crossed the Nari River five times, and pursued a tortuous course round the bends of the gorge; the Gundankinduff Defile, requiring the construction of two tunnels in treacherous material; the spectacular Chappar Rift, a chasm 4.3 km long through which the line had to pass by way of nine tunnels with a collective length of 1,981 m, a viaduct 23 m high and one large bridge 76 m above the stream below;[36] and the so-called 'Mud Gorge' on the summit, a narrow, winding and steeply sloping valley, 8 km long, with a shale bed consisting mostly of gypsum and clay. This material tended to swell and rise several feet after heavy rain, and then subside when it dried out. Because it was impossible to circumvent this gorge, a reinforced floating tunnel had to be drilled right through it.

Map 5 Sind Pishin and Bolan railways

Compounding all these difficulties were climatic and political challenges, brilliantly summarized by a civil engineer:

> The line does not wind its way through smiling valleys to the breezy heights above, and then, after a rush through an Alpine region, break out with mile upon mile of verdant plain, like the railways which lead from France and Germany to Northern Italy. On the contrary, it traverses a region of arid rock without a tree or bush, and with scarcely a blade of grass – a country on which nature has poured out all the climatic curses at its command. In summer the lowlands are literally the hottest corner of the earth's surface, the thermometer registering 124° Fahrenheit in the shade, while cholera rages, although there is neither swamp nor jungle to provide it a lurking place. In winter the upper passes are filled with snow, and the temperature falls to 18° below zero, rendering outdoor labour an impossibility. The few inhabitants that the region possesses are thieves by nature and cut-throats by profession, and regard a stranger like a gamekeeper does a hawk – something to be bagged at all costs. Food there is none, and water is often absent for miles; timber and fuel are unknown, and, in a word, desolation writ large is graven on the face of the land.[37]

Accordingly, the building had to proceed on a seasonal basis, addressing the lower sections of the route in the winter months until the onset of summer, whereupon everything was transferred to the higher and cooler levels. This would involve moving 15,000 to 20,000 workpeople as well as all their stores, tools and supplies and recommencing a new organization in a different place. Meanwhile Browne, who had expected unfettered control of the project with a remit to press on with all possible speed, found himself answerable to Terence Hope, ICS, a member of council for the Public Works Department. While the Royal Engineers left accounts of acrimonious exchanges between the two, especially over the financial estimates,[38] Hope toured the entire line in March 1884, and reported to the viceroy, Lord Ripon. Hope described the differing challenges of the sections, and commended the progress already made on the lower sections in making cuttings, embankments, one tunnel and nine pier foundations for the six of the bridges. There were still fourteen bridge gaps, and until the lower section was completed, 'even as a road', it would be impossible, he wrote, 'to make any use of the Hurnai [*sic*] route'. Girders and 64 km of rails had to be brought from England at a cost of £120,000 if they were going to complete those bridges. He assured Lord Ripon that 'everything is being done which is possible under the present arrangements', but in view of the Russian advances, he maintained that 'those arrangements are not adequate to the occasion'. He advised pushing on with the lower section of the route, so that in an emergency troops could at least move along that part of it.[39]

Once news had reached London of Russia's imminent annexation of Merv in February 1884, the secretary of state for India, the earl of Kimberley, met several cabinet colleagues, including Gladstone, to discuss this 'very grave occurrence, bringing as it does Russia into immediate contact with Herat territory'. They agreed that it might be 'expedient to proceed with the Quetta Railway',[40] a *volte-face* that several soldiers thought that might have embarrassed the Gladstone government.[41] Kimberley was unabashed: of the railway route, he added, 'I do not think we have really lost anything by the delay in making it. Had we persevered with it in 1880, we should have had firm opposition at home. Now, when the hot fit is coming again from our fickle public, it will be acquiesced in without demur.'[42] By making this announcement on 21 April 1884, he delighted Sir Donald Stewart, the commander-in-chief in India, who noted that 'we have now received authority to go on openly with the Quetta Railway'.[43] This would apply to both routes, and, in May 1884, he commented in his notebook, 'I am happy to say there is no longer any grumbling over the money spent on the Bolan Road and Harnai Railway.'[44] Meanwhile Roberts perceived the strategic and diplomatic potential of the railway. As he informed the duke of Cambridge, the railway to Quetta and Pishin 'is the first step towards strengthening our frontier in the Kandahar direction . . . We should do all in our power to attract the Afghans towards us.'[45]

Improvements in the shorter Bolan road had enabled Stewart to advise the duke of Cambridge that 'The road through the Bolan is progressing famously: it will not be equal to a Railway but it will be of incalculable value in a military sense to our frontier.'[46] Colonel Lindsay now had permission to build a railway through to Quetta,[47] and, in order to accelerate construction, the rails were laid mostly in the bed of the Bolan River, which was usually dry but was subject to flooding periodically. The first 64.4 km were fairly gentle but then the route moved on to a much steeper incline, precluding the use of a broad gauge. Hirok, some 80.5 km from Rindli and 1,371 m above sea level, became the changing station, whereupon 14.5 km of narrow gauge took the line over the summit of the pass, 1,707 m above sea level. Engines, carriages and wagons for the narrow-gauge section had to be brought up the mountains on the backs of elephants, images of which were reproduced around the world.[48] Thereafter another changing station enabled the 40 km of remaining route to reach Quetta over a broad gauge along an almost level plain.

Initially designated as a temporary line, the Bolan railway ensured that the first steam locomotive reached Quetta in August 1886. The railway did so only by drawing labourers away from the much more demanding work along the Sind Pishin State Railway. Eventually the Bolan railway required a thorough overhaul, following the floods of 1889 and the inconvenience experienced with the change of gauge. A complete reconstruction began in November 1891. The new project, known as the Mushkaf–Bolan railway, extended 138 km in length. It used some of the old alignment from Nari Bank, near Sibi and Kolpur, 40 km from Quetta, covered gradients varying from 1 in 55 to 1 in 25 and required eight tunnels and four crossings of the river up the Mushkaf Valley, three tunnels in the Bolan Valley and a very long tunnel from one valley to another. Near Kolpur the engineers built nine bridges over the Bolan River.[49]

Meanwhile the Sind Pishin Railway had made slower progress. During the summer of 1884 sickness, fever and scurvy took their toll as work concentrated on survey and tunnel work in the Chappar Rift (see Figure 5). In one gang of forty-two men, thirty died from scurvy, and everyone, as G. K. Scott Moncrieff recalled, was 'more or less saturated with fever'.[50] By October, as work moved to the lower reaches of the line, and under orders to press ahead and utilize rails, a brief outbreak of cholera precipitated the mass desertion of Pathans, who comprised the bulk of the labourers. Scott Moncrieff had to find several hundred masons and bricklayers from Delhi and the eastern Punjab as replacements, and further progress was made despite violent floods that swept away bridges and temporary roads. All these problems

Figure 5 The Sind Pishin Railway: viaduct over the Chappar Rift

recurred in even more devastating form in 1885. Rainfall that had been just over 5.8 cm in the first three months of 1883, and had become 12.4 cm in a similar period during 1884, increased to 48.9 cm in the first months of 1885 up to May. Causing widespread destruction up and down the line, the rainfall interrupted both communications and the lines of supply. Fever and cholera followed the deluge; many key engineers succumbed, and during June 1885 cholera claimed the lives of 2,000 men out of the 20,000 engaged on the works.[51]

Nevertheless, Browne persevered. He had the services of some extraordinarily resilient and enterprising engineers, including Captain (later Colonel Sir) Buchanan Scott, who worked in Gundankinduff Defile in the winter and on the Chappar Rift during the summer. Quite apart from all the difficulties of surveying the Rift, often undertaken in bare feet because 'the rocks', as Scott recalled, 'were too slippery to stand on',[52] and the 600 tons of iron girders that had to be brought up to erect the highest bridge, 88 m above the river bed below, there was the task of persuading workmen to start the adits (openings) on near-perpendicular cliff faces for several of the tunnels. These men had to be lowered in rope chairs from the cliff above, and then to use steel crowbars to bore holes about a foot deep into the cliff face at points previously marked with whitewash (from weighted bundles suspended at the desired lengths from ropes lowered from the cliff tops). Once holes were bored, and the crowbars inserted, platforms could be erected from which the adits were driven in and the blasting begun.[53] Not a man was lost on this part of the line, although the men were battling malaria in the autumn of 1885. Scott himself took seventy-five grains of quinine daily for seven days and found his digestion 'ruined' by the end of the week.[54] The year 1886 proved much easier, with a reduced incidence of flooding and sickness, and so construction proceeded apace. By 27 March 1887, an engine ran over the whole line from Sibi to Quetta, and a distinguished company, including the duke and duchess of Connaught and General Roberts, attended the opening of the bridge over the Chappar Rift. It was named the 'Louise Margaret Bridge' after the duchess, the first woman to see the bridge.

Roberts was certainly pleased. After a change of government in London, the new prime minister, Lord Salisbury, had assured him that 'the scientific frontier doctrine which was so much derided some years ago is master of the field now'.[55] Commenting in 1886, as the new commander-in-chief in India, upon 'Proposals for the Defence of the North-West Frontier, by the Defence Committee', Roberts urged that materials for an extension of the railway to Kandahar should be collected on a 'convenient point on the Bolan Railway', and that a couple of stern-wheeled Yarrow boats, as recently used on the Nile, should be sent up with British forces to the Helmand River.[56] By 8 May 1888, with both railways to Quetta operational, Roberts was delighted to see the railway being extended northwards to the frontier post of Chaman. He noted that current estimates envisaged the completion of the Khojak Tunnel (then the longest tunnel in India at 3.9 km) within two years, and expected that reserve material for the further extension to Kandahar could be stored at Chaman by the end of 1888. 'From the proposed terminus at Chaman', he observed, 'Kandahar is distant 78 miles [125.5 km].'[57] This entire route was enhanced in the following year by the erection of a splendid cantilever bridge at Sukkur, providing a direct crossing of the Indus River. The Khojak Tunnel (see Figure 6) was completed in 1891.

Figure 6 Khojak Tunnel

However spectacular the construction of the Quetta railways, they addressed only one of several routes by which Russia could invade India. Apart from Kandahar–Chaman–Quetta, the other two routes were from Ghazni through the Peiwar Pass and down the Kurram Valley to the Indus, and from Kabul through the Khyber Pass to Peshawar. The Khyber route received two surveys. The first concluded that an army operating beyond the Khyber could not be properly supplied because heavy traffic could not use the restricted space and steep gradients of the caravan road up the Khyber to Landi Kotal. The second survey, led by another redoubtable Scottish engineer, Captain (later Major-General Sir) James R. L. Macdonald,[58] reckoned that the Kabul River route would have to terminate near Dakka, Afghanistan, a politically impractical option.

Another survey, led by Buchanan Scott, examined the prospects for a railway in the Zhob Valley in 1890. This survey had to be escorted by an entire field force, under Major-General Sir George White, VC, which was restricted by the Indian government from venturing north of the Kundar River lest it alarm the Afghan regime. The field force moved in three columns through territory prone to tribal disturbances, partly to reward loyal Zhob chieftains and partly to overawe or pacify the Sheranis.[59] In other words, as fears of the Russian threat ebbed, other sources of instability on the north-west frontier became more acute. The signing of the Durand Line Agreement (12 November 1893) between Amir Abdur Rahman and Sir Mortimer Durand, a British diplomat, established an Indo-Afghan border that divided the Pashtuns (then known as Pathans), Baloch and other ethnic groups, who lived on both sides of the border. It enclosed the lands of the Chitral, Bajaur, Buner, Dir, Swat, the Khyber, Kurram and Waziristan within British territory, thereby augmenting the responsibilities of the Indian government while increasing the likelihood of conflict in the tribal regions.[60]

As British forces mounted expedition after expedition against the various hill tribes in the 1890s, the operational value of the northern railways became increasingly appreciated. The station at Nowshera on the line between Peshawar and Rawalpindi became a vital junction and railhead for several expeditions. When Major-General Sir Robert Low assembled his 15,000-man relief force, with followers and 28,000 pack animals, over a period of two and a half weeks for the Chitral expedition of 1895, Nowshera served as a base for the operation. In their acclaimed account of the expedition, the Younghusband brothers summarized the 'compendium of difficulties' that beset

> mobilization on the Indian frontier. Not only the troops and their stores have to be concentrated, but also many thousands of pack animals, and

> the food for the entire force, man and beast, for as long as the campaign lasts. Add to this that units had in some cases to come in immense distances, that the line of railway is a single one, and that the detraining station was a small roadside station without platforms or conveniences for disembarking troops, animals, and stores . . .
>
> It must be a source of gratification to the military that the scheme and the railway stood the severe tests applied to them.[61]

Nowshera was again the railhead for the ensuing expedition, the Malakand Field Force of 1897, which involved the services and demonstrated the literary credentials (in his first book) of Winston Churchill. As a young subaltern and war correspondent, he recalled travelling by rail for five days in 'the worst of the heat' from Bangalore to join the expedition. He observed that 'Those large leather-lined Indian railway carriages, deeply shuttered and blinded from the blistering sun and kept fairly cool by a circular wheel of wet straw which one turned from time to time, were well adapted to the local conditions.'[62]

Finally, the railway from Rawalpindi to Khushal Garh on the banks of the Indus proved an invaluable support for the largest operation of this period, the Tirah expedition of 1897–98. Writing from the base camp at Kohat, 51.5 km from the railhead, Colonel Henry D. Hutchinson described trains delivering '1500 to 2000 tons of stores daily' over several weeks, and the transport department 'moving this immense amount of supplies, clothing, and war material to Kohat on carts, camels, mules, ponies, and bullocks, an endless stream discharging on the stony plains around this cantonment'. 'Everyone', he noted, 'has an appointed task, and everything has an appointed place, and though upwards of 50,000 men and followers, and more than 20,000 animals, have to be daily provided for here, it is all done quietly and efficiently without fuss or delay . . .'.[63] Over a year later, he added, the railway was again crucial in the removal of the hundreds of sick and wounded, who had to be brought back from Bagh in the Tirah to the advance base at Shinauri (Shinkiari). This was a journey of 56 km by track, using mules, ponies, stretchers or doolies (litters), and under strong escort through hostile territory. Thereafter wheeled ambulances conveyed the casualties over 120.7 km to Khushal Garh, 'whence it was a run of eighty miles [128.8 km] to Rawal Pindi'.[64]

Railway developments in the North-West Frontier Province

A forward policy, involving hard-won and costly military victories, may have preserved political agencies under attack (as in the Chitral) and reaffirmed the military power and prestige of the Raj, but it had left challenging issues of consolidation. Thousands of British regular troops remained in the tribal areas lying between the Afghan frontier and the administrative border of British India. Lacking communications and easily isolated, these forces could be protected on a long-term basis only by the erection of expensive forts and the prospect of reinforcement. None of these options appealed to George Nathaniel Curzon, who had arrived as viceroy on 3 January 1899 and toured the frontier in the following year. He chose to defend the frontier areas by withdrawing British forces to the rear and replacing them with locally raised units like the Khyber Rifles and newer bodies which had been brought into being under British officers from the Indian Army. He also brought all the tribal areas together under a new administrative region, the North-West Frontier Province, with effect from 9 November 1901. A chief commissioner appointed by, and responsible to, the Government of India took charge, and Lieutenant-Colonel Harold A. Deane, the former political agent in the Malakand, was the first incumbent.[65]

Within this new security system, railways retained a prominent role, as the broad-gauge North Western Railway had just been extended from Peshawar to Jamrud, the so-called gateway to the Khyber Pass. Lord Kitchener, who became commander-in-chief in India in 1902, also toured the frontier within a year of his arrival and, as an engineer, exhibited a keen interest in extending the railway to the Khyber Pass. Of the two options, the Kabul River route was less challenging as an engineering prospect but could be easily threatened by Afridis firing from their bank of the river. The other route up the Loi Shilman Valley posed a real challenge for the engineers with a difficult drop down 305 m to Kam Dakka, and when the army sought approval for a survey it proved another source of friction between Kitchener and Curzon.[66]

Deane supported Kitchener with plans and estimates, but Kitchener's preference for a tunnel between Warsak and Smatsai was deemed too expensive and had to be set aside for an easier route round the bend in the river to Palosi.[67] Whichever route the railway took, it was liable to encounter resistance from local tribesmen, so engineers and officers of the Khyber Rifles met at Landi Kotal to discuss security arrangements. Major (later Lieutenant-Colonel Sir) George O. Roos-Keppel, a former political agent in the Kurram and Khyber areas, observed, 'The construction of the railway will undoubtedly be unpopular, and the line will be open to attack by the Mohmands from the north, the Afghans from the east, and the Afridis from the south.'[68] While his own regiment, the Khyber Rifles, could be expanded to provide guards, the only way to support both the Khyber Pass and the Kabul River railway was to establish outposts along the route, two to three days' march from their garrison. If the railway ran through the Loi Shilman Valley, the Khyber Rifles could protect the entire route only by acquiring six additional infantry companies with 100 sowars (cavalry). The more exposed Kabul River route would need nine additional infantry companies, with 100 sowars. The lines would have to be protected by blockhouses. On 6 July 1905 Deane informed the secretary to the Government of India that proposals had been developed to protect the 'working parties during the construction of the railway and for the safeguarding of the line after completion'. As the railway could face attack from the Afridis and other tribes on the right bank of the Kabul River, Deane reckoned that they had to engage the assistance of their historic enemies, the Mohmands, on the other bank. Accordingly, he did not recommend Roos-Keppel's proposal to use the Khyber Rifles to guard the entire railway but favoured using them only within the Khyber agency's area, so reducing the request for additional forces to 354 infantry and 80 sowars.[69]

After further investigations, construction work began on the Loi Shilman route, and, by 1907, 32 km of metre-gauge line ran from Khushal Garh to Jamrud, north along the Kabul River Valley and then westwards towards the Loi Shilman Valley. However, the signing of the Anglo-Russian entente in the Convention of St Petersburg (31 August 1907) removed any strategic rationale for the line because Russia declared Afghanistan to be 'outside the sphere of Russian influence'. Faced with mounting internal problems and the disastrous defeats of the Russo-Japanese War (1904–5), Russia had sought a rapprochement with Britain, and the ensuing treaty reduced the perceived threat to India. As the 'Great Game' ended, the strategic rationale for building military railways vanished, and, by 1909, the Kabul River railway was dismantled without ever reaching the Durand line.

Notes

1 R. A. Johnson, 'The Penjdeh Incident, 1885', *Archives*, vol. 24 (1999), pp. 28–48.

2 R. A. Johnson, '"Russians at the Gates of India"? Planning the Defense of India, 1885–1900', *Journal of Military History*, vol. 67, no. 3 (2003), pp. 697–743; A. L. Friedberg, *The Weary Titan: Britain and the Experience of Relative Decline 1885–1905* (Princeton: Princeton University Press, 1988), ch. 5; D. E. Omissi, *The Sepoy and the Raj: The Indian Army 1860–1940* (Basingstoke: Macmillan, 1994), p. 205; F. Kazemzedah, *Russia and Britain in Persia, 1864–1914* (New Haven, Conn.: Yale University Press, 1968), pp. 57–62, 72–8, 85.

3 P. Burroughs, 'Defence and Imperial Disunity' in Porter (ed.), *Oxford History of the British Empire*, vol. 3, pp. 320–45, at pp. 322 and 341.

4 A. Preston, 'Wolseley, the Khartoum Relief Expedition and the Defence of India, 1885–1900' in A. Preston and P. Dennis, *Swords and Covenants* (London: Croom Helm, 1976), pp. 89–122; I. F. W. Beckett, *Victorians at War* (London: Hambledon, 2003), pp. 45–52, and 'Soldiers, the Frontier and the Politics of Command in British India', *Small Wars and Insurgencies*, vol. 16 (2005), pp. 283–4.

5 Ultimately Roberts wanted the railway to reach Kandahar but this proved a political impossibility in his lifetime. Fearing that war was 'inevitable' in May 1885, he urged Sir Henry Rawlinson, then a Member of Council in India, that 'we should move an army corps onto Herat', Roberts to Rawlinson, 3 May 1885, in B. Robson (ed.), *Roberts in India: The Military Papers of Field Marshal Lord Roberts 1876–1893* (Stroud, Gloucestershire: Alan Sutton Publishing for the Army Records Society, 1993), pp. xvi, xviii and 319.
6 Kerr, *Building the Railways of the Raj*, p. 33.
7 Ibid. p. 1.
8 H. Bell, *Railway Policy of India* (Rivington: Percival, 1894), p. 46.
9 BL, India Office Records (IOR)/P/235/46B, 'Abstract of the Proceedings of the Hon'ble G. F. Edmonstone, Esq., Lieutenant Governor of the North Western Provinces in the Military Department', includes weekly reports from the escort to the Jubbulpore Railway Engineers from December 1859 through to 30 March 1860; see also Omissi, *Sepoy and the Raj*, p. 200.
10 BL, IOR/P/217/37, 'North West Provinces Lieut.-Governor's Proceedings in the Public Works Department (Railway)', reports nos. 25, 8 October 1862, and 83, 28 April 1864.
11 BL, IOR/P/191/15, 'Defences of Railway Stations', no. 57, A. J. Arbuthnot to secretary of Government of India, 21 April 1865.
12 Ibid. no. 58, Lt.-Col. C. H. Dickens, secretary to the Government of India, to the chief secretary of the Government of Madras, 16 June 1865.
13 Ibid.; when tensions eased, Lahore's design was modified: clocks were added to the look-out towers, and the fortified entrance was turned into a *porte cochère* with projecting canopies. Richards and MacKenzie, *The Railway Station*, pp. 68–70; Satow and Desmond, *Railways of the Raj*, p. 33.
14 J. Gaylor, *Sons of John Company: The Indian and Pakistan Armies 1903–91*, 2nd edition (Tunbridge Wells, Kent: Parapress Ltd., 1996), p. 32; see also Lt.-Col. A. A. Mains, 'The Auxiliary Force (India)', *Journal of the Society for Army Historical Research*, vol. 61, no. 247 (1983), pp. 160–85, and K. Roy, 'India' in Beckett (ed.), *Citizen Soldiers*, pp. 101–20, at p. 113.
15 BL, IOR, MSS EUR.F 108/24, White MSS, minute by Pritchard, enclosure no. 2 (29 August 1893), paragraph 22.
16 Gaylor, *Sons of John Company*, p. 32.
17 *Adaptation of Railways for Military Transport: Progress Report, 1877* (Simla: Government Press, 1877).
18 'Military Use of Railways in India: Precis of Report of the Railway Transport Committee, India, 1876', *Professional Papers of the Corps of Royal Engineers*, Occasional Papers, 2 (London: Royal Engineers Institute, 1878), pp. 22–39.
19 BL, IOR/L/MIL/7/5445, Report of the Commission chaired by the Hon. Sir Ashley Eden, 15 November 1879, pp. 35–6.
20 RA, VIC/ADD MSS E/12432, Cambridge MSS, Gen. Sir F. S. Roberts to duke of Cambridge, 7 June 1889.
21 Notably William P. Andrew, the chairman of the Scind and Punjab Railway; see his paper *The Indus and its Provinces, their Political and Commercial Importance Considered in Connexion with Improved Means of Communication* (London: W. H. Allen & Co., 1858), pp. 17, 31, 40.
22 S. Mahajan, *British Foreign Policy, 1874–1914: The Role of India* (London: Routledge, 2002), pp. 22–4, 60; Gen. J. J. McLeod Innes, *The Life and Times of General Sir James Browne, R.E., K.C.B., K.C.S.I. (Buster Browne)* (London: John Murray, 1905), p. 235.
23 M. Cowling, 'Lytton, the Cabinet, and the Russians, August to November 1878', *English Historical Review*, vol. 126 (1961), pp. 60–79.
24 This envoy was Sir Louis N. Cavagnari, who was later murdered with his escort, precipitating the second phase of the Anglo-Afghan War. B. Robson, *The Road to Kabul: The Second Afghan War 1878–1881* (London: Arms & Armour Press, 1996), ch. 7; Beckett, *Victorians at War*, ch. 5.
25 In addition Roberts wrote to many politicians and journalists, promoting his views that an Anglo-Russian war in central Asia was inevitable, and that India was

vulnerable unless the army seized the initiative and sought a forward defence along a 'scientific frontier' in Afghanistan: NAM, Add. MSS, 1971-01-23, Roberts MSS. For a description of these plans and how they evolved, see Johnson, 'Russians at the Gates', pp. 711–19.

26 Johnson, 'Russians at the Gates', pp. 705; Beckett, 'Soldiers, the Frontier', p. 284.

27 BL, Add. MSS 43565, fo. 218, Ripon MSS, Lord Ripon to secretary of state for India, 9 May 1880; see also BL, Add. MSS 43565, fo. 36, Ripon MSS, Lord Hartington to Ripon, 26 May 1880.

28 Locals found this hot desert location so miserable that they devised a rhyme about it: 'Since Sibi was created, what is the use of hell?' G. K. Scott Moncrieff, 'Sir James Browne and the Harnae Railway', *Blackwood's Edinburgh Magazine*, vol. 177 (1905), pp. 608–21, at p. 610; Porter, *History of the Royal Engineers*, vol. 2, p. 329.

29 The Harnae route was given preference because its maximum gradient was 1 in 45, with a maximum radius of curvature, 600 feet, as against 1 in 25 on the Bolan route, which could carry only lighter traffic and partly on a narrow gauge. Capt. G. K. Scott-Moncrieff, 'The Frontier Railways of India', *Professional Papers of the Corps of Royal Engineers*, vol. 11 (1885), pp. 213–56, at p. 219; McLeod Innes, *Life and Times*, p. 241; Porter, *History of the Royal Engineers*, vol. 2, p. 329; Sandes, *Military Engineer in India*, vol. 2, p. 145.

30 Scott-Moncrieff, 'Frontier Railways', pp. 218–19; see also L. Maxwell, *My God! – Maiwand* (London: Leo Cooper, 1979); P. S. A. Berridge, *Couplings to the Khyber: The Story of the North Western Railway* (Newton Abbot: David & Charles, 1969), pp. 135 and 140; N. Lera, 'The Baluchistan "White Elephant": The Chappar Rift and Other Strategic Railways on the Border of British India', *Asian Affairs*, vol. 31, no. 2 (2000), pp. 170–80.

31 In fifty years up to 1880 Russia advanced 2,000 km from the Caspian Sea towards Herat; in the next four years, ending in June 1884, Russia advanced another 1,000 km. By 13 February 1884, the Russians had annexed Merv, which was only 300 km from Herat. Mahajan, *British Foreign Policy*, p. 75. See also R. L. Greaves, *Persia and the Defence of India 1884–1892: A Study in the Foreign Policy of the Third Marquis of Salisbury* (London: Athlone Press, 1959), pp. 56–60, and Scott Moncrieff, 'Sir James Browne', p. 609.

32 BL, Add. MSS, 43523, fo. 40, Ripon MSS, earl of Kimberley to Lord Ripon, 14 February 1883.

33 Sandes, *Military Engineer in India*, vol. 2, p. 145 n. 2; Scott-Moncrieff, 'Frontier Railways', pp. 220–1.

34 Sandes, *Military Engineer in India*, vol. 2, pp. 143–4, 149–50; Scott Moncrieff, 'Sir James Browne', pp. 615–16; Maj.-Gen. Sir G. Scott-Moncrieff, 'Some Famous Engineer Officers of the Nineteenth Century', *RE Journal*, vol. 35, no. 3 (1922), pp. 113–28, at pp. 124–7.

35 Scott Moncrieff, 'Sir James Browne', p. 610; Sandes, *Military Engineer in India*, vol. 2, p. 145; Scott-Moncrieff, 'Frontier Railways', pp. 220–1.

36 Because the published heights of the bridge vary from 69 m to 88 m, I have used the measurement of 76 m (250 feet) given in the official history; see Porter, *History of the Royal Engineers*, vol. 2, pp. 330–1; Innes, *Life and Times*, pp. 243–5; Lera, 'Baluchistan "White Elephant"', pp. 173–4.

37 'The Sind Pishin Railways', *Engineering*, vol. 14 (1888), pp. 368–9, at p. 368.

38 Scott Moncrieff, 'Sir James Browne', pp. 614 and 616; Innes, *Life and Times*, p. 239.

39 BL, Add. MSS 43600, fos. 202–4, Ripon MSS, T. C. Hope to Lord Ripon, 31 March 1884.

40 BL, Add. MSS 43524, fos. 172–4, Ripon MSS, earl of Kimberley to Lord Ripon, 22 February 1884.

41 NAM, Add. no. 1971-03-23-97/1/LXXXV, fos. 156–8, Roberts MSS, Roberts to duke of Cambridge, 6 March 1884; Scott-Moncrieff, 'Frontier Railways', p. 220.

42 BL, Add. MSS 43525, fo. 7, Ripon MSS, earl of Kimberley to Lord Ripon, 14 March 1884.

43 Elsmie (ed.), *Field-Marshal Sir Donald Stewart*, note books, 21 April 1884, p. 414.
44 Ibid. note books, May 1884, p. 414.
45 NAM, Add. no. 1971-01-23-97/1/XCIX, fos. 180–1, Roberts MSS, Roberts to duke of Cambridge, 11 May 1884.
46 RA, VIC/ADD MSS E/1/10371, Cambridge MSS, Sir D. Stewart to duke of Cambridge, 2 April 1883.
47 After an accident in 1886 Lindsay had to relinquish his command to a civilian engineer, Mr (later Sir) Francis O'Callaghan, Sandes, *Military Engineer in India*, vol. 2, p. 145.
48 The article appeared in the French magazine *L'Illustration* and then as 'An English Military Railway', *Scientific American*, vol. 53 (15 August 1885), p. 105.
49 Scott-Moncrieff, 'Frontier Railways', pp. 239–40; Sandes, *Military Engineer in India*, vol. 2, p. 146.
50 Scott Moncrieff, 'Sir James Browne', p. 617.
51 Ibid. pp. 617–21; Innes, *Life and Times*, pp. 254 and 256.
52 Sandes, *Military Engineer in India*, vol. 2, p. 151.
53 Ibid.; Innes, *Life and Times*, pp. 245 and 256; Porter, *History of the Royal Engineers*, vol. 2, p. 331.
54 Sandes, *Military Engineer in India*, vol. 2, p. 152.
55 NAM, Add. no. 1971-01-23-80, fo. 4, Roberts MSS, Lord Salisbury to Roberts, 6 July 1885.
56 NAM, Add. no. 1971-01-23-95/VII, fo. 53, Roberts MSS, General Sir F. Roberts, 'Notes on "Proposals for the Defence of the North-West Frontier", by the Defence Committee in 1885', 22 June 1886.
57 NAM, Add. no. 1971-01-23-95/1/XXI, fo. 127, Roberts MSS, Gen. Sir F. Roberts, 'Note on the Defence Works of the North-West Frontier', 8 May 1888.
58 Major-General Sir James R. L. Macdonald, RE (1862–1927), was a balloonist, engineer, explorer and cartographer. He surveyed railways in India and East Africa, explored the Upper Nile region and led a major expedition into Tibet in 1903–4. For this survey, see Brig.-Gen. H. H. Austin, 'The Kabul River Survey', *RE Journal*, vol. 41 (1927), pp. 414–26.
59 Capt. C. McFall, *With the Zhob Field Force 1890* (London: William Heinemann, 1895), pp. 14, 102, 114.
60 Sir W. K. Fraser-Tytler, *Afghanistan: A Study of Political Developments in Central and Southern Asia*, 3rd revised edition (London: Oxford University Press, 1967), p. 172; see also M. Barthorp, *Afghan Wars and the North-West Frontier 1839–1947* (London: Cassell, 1982), pp. 99–100.
61 Capts. G. J. and F. E. Younghusband, *The Relief of Chitral* (London: Macmillan, 1895), pp. 60–1.
62 W. S. Churchill, *My Early Life: A Roving Commission* (London: Odhams Press, 1930), pp. 121–2.
63 Col. H. D. Hutchinson, *The Campaign in Tirah 1897–1898: An Account of the Expedition against the Orakzais and Afridis under General Sir William Lockhart. Based (by Permission) on Letters Contributed to 'The Times'* (London: Macmillan, 1898), pp. 48–9.
64 Ibid. pp. 242–3.
65 D. Gilmour, *Curzon* (London: John Murray, 1994), pp. 196–8.
66 TNA, PRO 30/57/29, fo. 31, Kitchener MSS, Lord Kitchener to Lord Roberts, 12 January 1905; see also Field Marshal Lord Birdwood, *Khaki and Gown: An Autobiography* (London: Ward, Lock & Co., 1941), pp. 145–6.
67 TNA, WO 106/56, Lt.-Col. H. A. Deane to the secretary to the Government of India, 6 July 1905, refers to the previous letter of 17 December 1904.
68 Ibid. Maj. G. O. Roos-Keppel to Maj. W. E. Venour, 31 May 1905.
69 Ibid. Deane to the secretary to the Government of India, 6 July 1905, and, on raising a levy corps of Mohmands and moderately increasing the Khyber Rifles, E. H. S. Clarke (deputy secretary to the Government of India) to the Hon. M. F. O'Dwyer, 28 August 1908.

6

The Sudan Military Railway

Using railways for operational support was the primary mission envisaged by the late Victorian army. Only a couple of the railways had been built completely in theatre during a conflict; both of these (in the Crimea and Abyssinia) were relatively short, and the latter was relatively far to the rear. None of the Victorian works of construction emulated the length and significance of the Sudan Military Railway, a remarkable undertaking that not merely enhanced the logistical arrangements but also altered the entire calculus of the campaign. It enabled its creator, Major-General Sir Horatio Herbert Kitchener, to bring his Anglo-Egyptian army of 23,000 men into the heart of the Sudan, remarkably fit and supported by artillery on the land and river. It was able thereby to achieve a decisive victory at Omdurman (2 September 1898) against a vast indigenous army (possibly 52,000 men), fighting under Khalifa Abdullahi on his chosen field of battle.[1] Brilliantly described by the war correspondent G. W. Steevens as 'the deadliest weapon that Britain has ever used against Mahdism', the railway, Steevens added, reflected 'all the differences between the extempore, amateur shambles of Wolseley's campaign and the machine-like precision of Kitchener's'.[2] If this description overlooks Kitchener's unpredictable temperament, secretive and autocratic working methods and abrasive handling of his staff, it recognises Kitchener's boundless energy, fierce determination and sense of mission,[3] which found their most expressive outlet in the conceptual design, construction and creative usage of the Sudan Military Railway (see Map 6).

Dongola campaign (1896)

In 1896, Kitchener had little reason to expect a campaign of reconquest in the Sudan. The Conservative prime minister, Robert Gascoyne-Cecil, the third marquess of Salisbury, may have wanted to avenge the death of the Christian hero Charles 'Chinese' Gordon,[4] repress the depravations of the Mahdist regime,[5] and secure the southern border of Egypt, but he was not tempted by any of these concerns into a rash

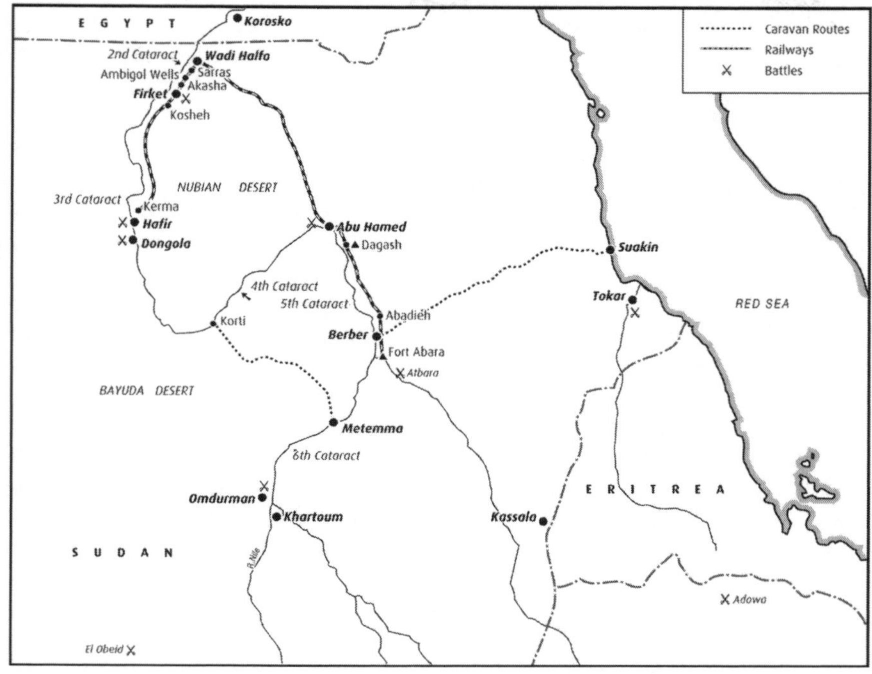

Map 6 Sudan Military Railway

move. It required pleas from Italy for some diversionary action by Britain in the Sudan to relieve pressure on its garrison at Kassala after the catastrophic defeat of Italian forces at Adowa (1 March 1896) before the cabinet resolved on 12 March 1896 to launch a limited advance into the province of Dongola.[6] At this time, there was neither any plan for an advance nor conviction that it need press on towards the town of Dongola, nor even any readiness to employ British or Indian soldiers.[7] Lord Cromer, the British agent and consul-general in Cairo, had raised the issue of employing British forces from the outset, and Lord Lansdowne, the secretary of state for war, had offered the services of the 1st Battalion, North Staffordshire regiment, 'as some slight support to the Egyptian troops' at Wadi Halfa on the Egyptian–Sudanese border.[8] Cromer, who was also anxious about the 'financial side of the question', expressed complete confidence in the prospective command of the 'cool and sensible' Kitchener, who 'knows his subject thoroughly – and is not at all inclined to be rash'.[9]

Kitchener had spent some fourteen years in Egypt, reforming the Egyptian army and fighting Mahdists on the Sudanese frontier. In 1892 he was appointed sirdar (commander-in-chief) of the Egyptian army and had some highly experienced assistants: Brevet-Colonel (later General Sir) H. M. Leslie Rundle as his chief of staff; Brevet-Colonel (later General Sir) Archibald Hunter, commandant of the Frontier Field Force and subsequently in command of the Egyptian Division; and Major (later General Sir) F. Reginald Wingate as his chief of intelligence. Wingate had served in Egyptian intelligence since 1887 and, as director of intelligence since 1889, had employed a small staff of skilful interrogators and intelligence analysts to collate information from an extraordinary network of spies, disaffected chiefs, defectors and former escapees, such as Rudolf Slatin and Father Ohrwalder. He provided Kitchener with detailed information on what was happening inside the enemy's lines, whether along the Nile or in the khalifa's innermost counsels.[10]

Finally, Kitchener obtained the indispensable services of Lieutenant E. Percy C. Girouard, a French Canadian, who would serve as his director of railways at the age of twenty-nine years. Having learned his craft on the Canadian Pacific Railway, Girouard was an unconventional officer who earned the confidence of Kitchener, lacked any deference towards senior officers and remained fiercely loyal to, and protective of, his subordinates. Once he joined the Dongola campaign at the end of March 1896, Girouard impressed Lieutenant H. L. Pritchard, RE, as knowing 'exactly what was wanted, and had the head, the energy, and the pluck to do it. He was here, there, and everywhere, instructing, swearing, shoving things along, gradually getting organization and system to take the place of chaos and under his able guidance every one became slowly, but noticeably, more proficient at the work . . .'.[11]

Initially, Egyptian forces composed the invading army while some 2,500 Indian soldiers were sent to garrison Suakin, and the North Staffordshires remained in reserve until the final advance on Dongola. When operating across a barren and desolate terrain bereft of roads, 'one of the main difficulties', wrote Cromer, 'was how to transport the food and stores for the army whilst on the march to Dongola'.[12] As camels were too slow and costly, and river transport could be used only between the rapids in the Nile, the railway between Wadi Halfa and Sarras had to be extended in a southerly direction. Accordingly, Akasha some 80 km south of Sarras became the first objective, and it was occupied without resistance on 20 March 1896.[13] Work on the railway, at first under Lieutenant A. G. Stevenson, RE, and later under Girouard, commenced immediately. As Colonel Sandes perceptively observed, 'The Dongola Expedition, and the Omdurman Campaign which followed it, consisted of building railways with attendant military operations rather than of military operations with attendant railway construction.'[14]

Extending the railway involved clearing the desolate village of Akasha, establishing telegraphic communications with Wadi Halfa, building a new fort and preparing an entrenched camp. Thereafter three battalions of infantry, some cavalry, a camel corps and a battery of artillery protected the railway battalion, which employed conscripted *fellahin* as it repaired the torn-up line and built new track. By the end of the month they had rebuilt the line as far as Ambigol Wells, about 103 km from Wadi Halfa and only 37 km from Akasha. As the daytime temperatures soared above 116° Fahrenheit in the shade, often to over 120° and sometimes to 127° and 129°, the labouring crews often worked through the night.[15] Meanwhile Kitchener had steadily concentrated his forces between Sarras and Akasha, with over 8,000 deployed by 12 April. The khalifa had rushed forces northwards, and, on 1 May, in a skirmish to the east of Akasha, the Mahdists were driven off into the desert. By early June, the railway was within striking distance of the main concentration of Mahdists at Firket (Firka), and camels supplied the Egyptian troops as they advanced from the railhead to dislodge the enemy in a surprise assault (7 June 1896). Cromer was delighted that Kitchener's plan had succeeded 'admirably', even 'to the exact date (June 7th) which he had stated weeks beforehand . . . the Egyptians fought under the most favourable conditions possible. The result was a complete success.'[16] Wingate, who had provided remarkably accurate information about the number and disposition of the Mahdists at Firket, doubted that 'we shall have much trouble whilst waiting for railway and steamers & unless strong Dervish reinforcements come, the final reoccupation of the Province ought not to be a matter of difficulty'.[17]

The railway and river decided the pace and tempo of subsequent operations. Kitchener's forces had to spend a fortnight at Firket, waiting for the completion of the railway to this site and the rising of the river, which would allow boats to be brought up the cataract and sail southwards. Firket, however, soon became insanitary owing to the numbers of inadequately buried corpses, and so the entire force moved on to an advance base at Kosheh (Kosha), 174 km from Wadi Halfa. While the railway crept steadily onwards, reaching Kosheh on 4 August, the expedition had had to cope with its first major challenge, an outbreak of cholera that reached Lower Egypt from Cairo in late June. The disease swept southwards along the railway and camel convoys, and it was soon coupled with an outbreak of enteric fever (typhoid fever). Casualties mounted rapidly despite efforts to spread out the military camps and minimize intercourse between them. The official death toll from disease, as presented to the House of Commons, was 235 soldiers (compared with 47 killed in action), but subsequent accounts claimed that 260 Egyptian soldiers and 640 camp followers had died from disease, as well as 19 British soldiers.[18] By 13 August, Major John Francis Burn-Murdoch, formerly of the Royal Dragoons and now an officer with the Egyptian cavalry, recorded in his diary that they had not had any cholera 'for nearly two weeks so may be said to be clear of it. On the whole we have got through it very well.'[19]

Having surmounted this challenge, Kitchener was informed on 28 August that freak winds and rain had washed away 19 km of railway between Wadi Halfa and Akasha, and that it might take three weeks to repair. Fearing lest the river would fall again after such a delay, Kitchener assigned three and a half thousand labourers, working day and night, to the task, and he not only supervised the repair but also participated in the labouring work. By 6 September 1896, the crew had completed the task.[20]

Kitchener was now able to send forward his fully supplied Dongola Expeditionary Force, an Anglo-Egyptian body of 15,000 men with four batteries of artillery, a Maxim battery and a large camel corps. It had, too, substantial support on the river, including four gunboats and three unarmed steamers but not the pride of the flotilla, the *Zafir*, which had arrived by rail in prefabricated sections. Designed by Kitchener and built in England, it had been trans-shipped seven times and had travelled 6,437 km before its arrival in Kosheh just before sunset on 14 August. Mightily impressed, E. F. James of *The Times* described 'a train with four trucks bearing what appeared to be four huge square iron cases, painted red, each as big as a two-storeyed house and towering high above the engine'.[21] These were the first four sections of a ten-section, stern paddle-wheel vessel, with three decks and capable of carrying a 12-pounder gun, two 6-pounders, a small howitzer, twelve Maxims and a searchlight. Although the deaths of both its chief engineer and second engineer handicapped the reassembly of the *Zafir*, it appeared ready to join the renewed offensive on 11 September. Once Kitchener had boarded the vessel, and spectators crowded the riverbanks, the *Zafir* began its maiden voyage, only for a cylinder to burst and render it out of action.[22] Ironically, the ensuing and decisive battle at Hafir (19 September 1896) became an extended bombardment between the artillery and gunboats on one side of the river and the Mahdists on the other side. It took over three hours to silence the Mahdist guns and enable the gunboats to steam on towards the town of Dongola. Meanwhile at Hafir the Mahdists, led by Wad Bishara, who found themselves outnumbered by three to one and outflanked on the river, prudently evacuated the town. Four days later the gunboats, including the reconditioned *Zafir*, bombarded the defences of Dongola, and after some desultory street fighting, the town surrendered. The cavalry harried the retreating Mahdists and brought back some 900 prisoners.[23]

At a victory parade on 24 September Kitchener praised the contribution of all his forces, including the North Staffordshires, who had joined in the final assault. He apologized to the latter for not giving them much of a battle but commended their 'valuable assistance in mending the railway and pulling the boats up'.[24] Nor was it only Kitchener, the engineer, who appreciated the primacy of logistics for colonial campaigns: so did Colonel Hunter, sometimes known as 'Kitchener's sword arm'. As the dashing colonel wrote during the midst of the Dongola expedition:

> More than two-thirds of the work is calculating the quantity of supplies required and where to have them and by what time. In fact, war is not fighting and patrolling and bullets and knocks; it is one constant worry about transport and food and forage and ammunition and seeing that no one is short of stuff and that collisions don't take place on a single line of railway.[25]

Although Kitchener was aware of the risks of leaving a small Egyptian force in hostile territory at the end of a single railhead, albeit one that still advanced southwards until it reached Kerma in May 1897,[26] Cromer had preferred that the force should remain in Dongola for several years.[27] He was persuaded by Kitchener among others that the military successes had been more easily and cheaply attained than expected; 'that the dervishes should not be given time to recover'; and that Kitchener, if funded with £500,000, could build 'a railway from Korosko to Abu Hamed', purchase more gunboats and even occupy Berber.[28] Kitchener took his case to England, and the prime minister was astonished at the outcome: as he informed Cromer, 'His campaign against the Chancellor of the Exchequer was not the least brilliant and certainly the most unexpected of all his triumphs. But all his strategy is of a piece – the position was carried by a forced march and a surprise.'[29]

Building the Sudan Military Railway (1897)

Buttressed now with funding and political approval, Kitchener could plan his next railway. He had already jettisoned the options of crossing the Bayuda Desert, and of advancing from Suakin, both of which were associated with the failures of the mid-1880s. Instead he decided to make a fresh start by crossing the Nubian Desert to Abu Hamed.

He also determined to employ the broad gauge of 3 feet 6 inches, and not the narrower one of 3 feet 3 inches favoured by Cromer, and did so because it matched the gauge of the railway being built in South Africa by Cecil Rhodes, the much-vaunted Cape–Cairo railroad. This had an immediate advantage when Girouard returned to England to purchase fifteen new engines and 200 trucks inasmuch as Rhodes was willing to lend him some 70-ton and 80-ton locomotives previously destined for the Cape and Natal railways.[30]

Construction of the Desert Railway began on New Year's Day 1897, but progress was initially quite slow while workshops were being built in Wadi Halfa, 1,500 platelayers were being trained, and rolling stock was being brought from England. Only 64 km had been laid by May, when the line to Kerma was completed and officers and men could be switched to the desert line, and new engines employed to support the track laying. By 5 June, Kitchener reported that 'The desert line is going well 63 miles [101 km] laid. My railway battalion laid the other day 1,900 yards [1,737 m] in 4 hours & 10 minutes so you see they are pretty smart & can do 2 miles [3.2 km] a day if we could only get the material up . . .'.[31] Although the railway crew once managed over 4,572 m in twenty-four hours,[32] their average rate, as Kitchener claimed, was about '1½ to 2 miles [2.4 to 3.2 km] a day provided the earthwork is not very severe'.[33]

None of this data, though, concealed the extraordinary vulnerability of the railhead as it pressed forward across a desolate and barren landscape under the ceaseless pressure of a tropical sun. The canvas town of 2,500 inhabitants, complete with station, stores, canteen, post office, telegraph office and a reserve of 10,000 gallons of water, depended entirely upon a single railway track to survive and function. Every morning a 'material' train arrived, as Churchill described, 'carrying its own water and 2,500 yards [2,286 m] of rails, sleepers, and accessories'. At noon a supply train followed, 'also carrying its own water, food and water for the half-battalion of the escort, and the 2,000 artificers and platelayers, and the letters, newspapers, sausages, jam, whisky, soda-water, and cigarettes which enable the Briton to conquer the world without discomfort'.[34] The work was grindingly monotonous: every day platelaying gangs unloaded material from the incoming train and laid sleepers in a long succession, spiking every alternate one, and then the 80-ton 'camp-engine' moved cautiously forward along the unballasted track, followed by more platelayers, who completed the spiking and ballasting process. The grinding pace reflected the perception that the Nubian Desert was waterless but, in the middle of May, at what became No. 4 station, 124 km from Wadi Halfa, Kitchener ordered the digging of a well. After five weeks of digging, abundant

Figure 7 'Gangs of Rail-Layers Leaving by Train for the Railhead', *Black and White War Albums. Soudan No. 2: Atbara. Snapshots by René Bull, Special Correspondent to Black & White* (London: Black and White Publishing Company Ltd., c. 1898)

supplies of water were found at a depth of 22 m, and with the use of a steam pump, the well yielded a continual supply. Kitchener's 'luck' held again in October, when a second well was dug at No. 6 station and even more water was found at a depth of 27 m. If the wells did not solve the water problem entirely, they greatly enhanced the carrying capacity of the line and reduced the danger to which the labourers (see Figure 7) were exposed.[35]

The line, nonetheless, was being laid at great speed, within a tight budget and across hostile terrain. By 23 July, the railway had reached its summit, 166 km from Wadi Halfa, without ever exceeding the limits of curve and gradient prescribed by Girouard (namely a six-degree curve and a 1 in 120 gradient). It was now too dangerous to proceed further without securing the terminus at Abu Hamed. Hunter's capture of that place on 7 August, after marching his column from Merowe along the river bank, followed by the arrival of five gunboats on 29 August, reduced the chance of any attacks on the line or the workforce. The khalifa, who had never seen a railway, failed to comprehend its significance and never sent forces north to impede its development (and probably lacked the supply and transport capacities to do so). He concentrated instead upon the caravan route from Korti to Metemma across the Bayuda Desert, which the invaders had used in the mid-1880s. Accordingly, he gave priority to quelling a rebellion by the Jaalin tribe in Metemma. Kitchener rejoiced that 'The Dervishes are divided amongst themselves & the Khalifa is having trouble . . . I am in communication with a good many in the interior who desire to upset his rule.'[36]

Kitchener's anxieties soon revived when he received news that a patrol of forty soldiers on camels, having advanced from Abu Hamed, had occupied Berber on 31 August. They had found the town of 12,000 inhabitants evacuated by the local emir, Zaki Osman, and his troops had fled, apparently fearing attack from the sirdar's gunboats. However, Berber was another 209 km up river, and it was vulnerable to attacks on its line of communications or even recapture from the large Mahdist forces under Emir Mahmud which were now ensconced in Metemma. Although Kitchener knew that this move on Berber was ahead of schedule, not least because he was still battling for resources from the finance department in Cairo, he despatched Hunter with nine British officers and 350 men from the 9th Sudanese Battalion to garrison Berber and reopen the Berber–Suakin road. Kitchener later visited Berber in the middle of September and promptly authorized the despatch of reinforcements, including the rest of the Sudanese brigade, field artillery and two camel corps companies. He also sent a telegraph detachment to Berber, with the polling party often 241 km behind the wire-laying party, and gunboats up the Nile to deter Mahmud by shelling his camp and forts outside Metemma.[37]

Only a secure line of communications could now sustain Berber, but the Nile route would not suffice. As soon as the waters began to fall, the cataracts south of Abu Hamed proved, in Kitchener's words, 'far worse' than anticipated. At that time he had a mere 27 km of rail to stretch beyond Abu Hamed and so feared that 'it will take the best part of a year to extend the railway to Berber'.[38] Still battling with Cairo for resources, and profoundly stressed, Kitchener even threatened to resign.[39] Once assured of Cromer's continuing support, Kitchener withdrew his threat and was able to make more realistic plans for the railway, which had reached Abu Hamed by 31 October. By 19 November, he assured Sir Evelyn Wood that

> We must if possible get the railway through to Berber before anything is done. I hope to open the line early in April. I have had to fight for money but I have got enough to keep the Army in a state of efficiency in its present positions . . .[40]

Strategically, the railway with its accompanying telegraph had already changed the calculus of the campaign. It had shortened the time of the 370-km journey from Wadi Halfa to Abu Hamed from eighteen days by camel and steamer to twenty-four hours (depending upon the serviceability of the engines; see Figure 8). It had also enabled Kitchener to move his forces into the heart of the Sudan, independent of season or of the height of the Nile.[41] An extraordinary network now existed from Cairo to Abu Hamed, whereby stores travelled by train for 620 km to Naghamadi; thence by boat some 233 km to Assouan; then by train round the cataract to Shellal, another 6.4 km; and finally by boats lashed alongside sternwheelers that were towed for 386 km up to Wadi Halfa. Supervised by British officers, convict labour undertook the loading and unloading.[42]

Apart from the use of the existing materials to extend the Desert Railway another 26 km to Dagash, railway construction paused in the last couple of months of 1897. Supplies were brought forward to transform Abu Hamed from an isolated outpost into a major base. The falling of the river also revealed an unexpected cataract 6.4 km

Figure 8 'Air-Condensing Engine Crossing the Desert', *Black and White War Albums. Soudan No. 2: Atbara. Snapshots by René Bull, Special Correspondent to Black & White* (London: Black and White Publishing Company Ltd., c. 1898)

north of the confluence of the Nile and Atbara Rivers, and Kitchener, rather than withdraw the gunboats north of this obstacle, authorized the construction of a small dockyard at the confluence. This enabled the gunboats to continue their reconnaissance missions further south. Although the route of any future campaign was now taking shape, neither Lord Salisbury nor Lord Cromer nor even Kitchener wanted to bring forth British reinforcements. As late as 25 December 1897 Kitchener proffered reassurances that Mahmud's forces lacked the supplies to move north, and that only raids were likely before the railway could be extended to Atbara. Deploying British forces at this stage, he argued, could leave the soldiers in an insanitary base such as Wadi Halfa, while hampering 'transport on the railway' that should 'be fully engaged in getting up railway material'.[43]

Meanwhile Cromer, writing on the same day, informed Lord Salisbury that 'whether we like it or not, an expedition to Khartoum next year will be almost unavoidable'. He did not like 'the idea of an English Expedition' but doubted that the Egyptian army, 'which is neither very strong in point of numbers or very trustworthy', could hold all of its extended positions deep in the Sudan. His sole consolation was the prospect of an 'interval before the question of an Expedition is ripe for solution'.[44] This confidence was promptly blown away by Wingate's intelligence that the khalifa had planted his standards at Karari 11 km to the north of Omdurman, seemingly intent on concentrating a large force for a possible move northwards. The evidence, as Wingate confided in his diary, 'may or may not be true but I thought it quite good enough to urge the Sirdar to call now for British troops and after a long discussion he at last agreed and the monstrous and important step was taken'.[45] Because apparent confirmation of these rumours followed quickly,[46] Cromer endorsed the request, and the lead elements of the first of two British brigades were despatched from Cairo.

Omdurman campaign (1898)

In the event the intelligence proved mistaken, but the die was cast and plans for the campaign accelerated. As additional Egyptian forces were sent from Dongola and Suakin to Berber, the 1st battalions of the Lincolnshires, Royal Warwickshires and Cameron Highlanders rushed to camps near Berber, followed by the Seaforth Highlanders in March. This concentration of a powerful army of about 14,000 men required the maintenance of a complex and costly supply chain, relying on boats and camels. Girouard, however, was able to resume his railway building in January along the bank of the Nile, with expert crews working in relatively cool weather. By 10 March, the railhead was 121 km from Abu Hamed, and because reconnaissances indicated that the Nile was broken and rocky for the next 80.5 km, Girouard detoured the line through level desert before bringing it back to the Nile bank just above the cataracts near Abadieh on 5 May. He took the line steadily southwards, reaching Berber after another 35 km, and thence on to the camp at Darmali before finally reaching Atbara – 250 km from Abu Hamed and 620 km from Wadi Halfa – on 3 July 1898. As Churchill prophetically observed, 'On the day that the first troop train steamed into the fortified camp at the confluence of the Nile and the Atbara rivers the doom of the Dervishes was sealed.'[47]

Ever since the line had reached Abadieh, the railway enabled new gunboats – *Melik*, *Sultan* and *Sheikh* – to be brought from Wadi Halfa and reassembled at reception sidings on the bank of the Nile. As soon as the railway reached its new terminus, Fort Atbara, supplies poured in; two trains, bringing some 350 to 400 tons between them, arrived daily.[48] By 17 July 1898, Kitchener could report that 'The railway is now practically finished and the bridges are nearly all done. I am running supplies on as quickly as possible and if we get good wind this month everything ought to be ready at Atbara in time.'[49] In due course he brought the rest of his army forward, including a flotilla of ten gunboats (one of which, the *Zafir*, would sink) and five steamers on the river; as well as the 4th Egyptian and 2nd British Brigades, thereby assembling an army of 23,000 men, with forty-four guns, twenty machine guns and cavalry, fully equipped and supplied for the final advance on Omdurman. By transporting and maintaining this force over 1,609 km from its main base in Cairo, Kitchener and his engineers had achieved 'a remarkable feat of engineering and organization'.[50]

Many different experiences occurred on the journey from Egypt. The railway stations at Cairo and Alexandria provided focal points for British residents to come and cheer the departing soldiery. These were morale-boosting experiences, as Captain Samuel Fitzgibbon Cox (Lincolnshires) recalled: 'The station' at Cairo, he wrote, was 'absolutely packed with the elite of society much weeping & gnashing of teeth. 21st officers handing round cigars & champagne in silver cups . . . personally never knew how damn popular I was before!!'[51] The Warwickshires received a similar reception when they left Alexandria, but thereafter the travelling experiences of officers and other ranks differed quite sharply. Second Lieutenant H. P. Creagh-Osborne (Royal Warwickshires) described a 'very comfortable' journey, with the train stopping 'every hour or so at stations in the open desert', while Drummer W. W. Taylor found that 'We could not sleep in the train as there was no room to lie down. In fact we could scarcely sit down, being packed in like sardines . . .'.[52] Many soldiers remembered being packed into trucks, sometimes forty-five or forty-seven per truck, and standing all the way as they travelled for thirty-six hours or sometimes for two days on the Sudan Military Railway. As one Rifleman reflected bitterly, 'we had to stand up nearly all the time, so tightly were we packed in them. We lived on corned beef and biscuits nearly all the time . . .'.[53] In this respect the taller Grenadier Guards had an advantage, as only thirty-five Guardsmen travelled in each truck, but the Guardsmen complained about a dreadful journey through Lower Egypt. Ensign G. D. Jeffreys (Grenadier Guards) remembered 'an awful night & morning in the train' to Luxor and then a steamer that was 'comparatively comfortable though of course v. crowded, & one got a little air, after the awful heat & could sleep on deck at night'.[54] In travelling to assume command of the 2nd British Brigade, Brigadier-General (later General Sir) Neville Lyttelton had the misfortune to find himself stranded in the Nubian Desert when his engine driver fell ill. In waiting eight hours for another train, he was surprised by the lack of dust but found the glare very 'bad' and the thirst a real strain.[55]

The time-consuming burden of loading and unloading varied with the corps, but it was particularly daunting for those in charge of the guns. Captain D. W. Churcher (1st Battalion, Royal Irish Rifles), though in charge of the machine gun battery, also found himself responsible for the transport of the howitzers and 40-pounder guns. Accordingly, at Wadi Halfa, he had his men turned out at 3.00 in the morning to commence unloading the guns from the vessels and then load them onto the first train. The process took four hours, whereupon they had breakfast in the Egyptian army mess and then began loading the second train, which had 'the men of the Howitzer Battery, the 40-pounders and my own and all the mules, horses and heavy baggage, so it's a pretty long one'.[56]

'H.F.' (Major Harry Finn, 21st Lancers) described the four-stage journey of the cavalry: initially twenty-four hours by train from Cairo, 'very hot and dusty, and attended with considerable difficulty in watering the horses'; then being towed down the Nile, followed by a canter round Shellal cataract that 'delighted' the horses and another train journey with 'coolies and convicts' loading and unloading; sternwheelers down to Wadi Halfa; and thence on to the thirty-six-hour journey by train to Atbara. He recalled waking to see that 'we were in the middle of the desert, not a living thing to be seen, simply a waste of smooth sand with an occasional rocky hillock, black and barren'.[57] The railway had left its mark on this landscape: the stations designated No. 1, No.2 and so on were simply mud huts, as Lieutenant

Ronald F. Meiklejohn (Royal Warwickshires) recalled, and the track had its own imagery as captured by Captain Churcher:

> We are just about the middle of the Nubian Desert and a more desolate looking spot I have never seen, it is all sand, flat as a pancake with the dim outline of hills East and West of the line. There is not a tree or a blade of grass anywhere, nothing but sand, the railway track marked by sardine tins and broken bottles.[58]

The railway certainly had one disadvantage inasmuch as it drastically reduced the period of acclimatization. This was a particular problem for the 2nd British Brigade, which was transported much faster than its predecessor (by means of extra rail links from Cairo to Assouan and from Abu Hamed to Atbara), and, in early August, into the heart of the Sudan. After a seven-day journey Sergeant-Major Clement Riding (Royal Army Medical Corps) found the heat 'something cruel' on his arrival at Fort Atbara.[59] However, any grumbling about conditions on the railway paled into insignificance for the soldiers of the 1st Brigade, as Major-General William F. Gatacre had ordered them to march from Abu Hamed to south of Berber to check a reported move by the Mahdists. Hundreds of Lincolns, Warwicks and Cameron Highlanders fell out on the line of march through deep sand and without any shade. This proved one of the more exhausting ordeals of the entire campaign, made worse by the subsequent realization that the threat to Berber 'was mainly Gatacre's imagination'.[60] Conversely, the wounded benefited immensely from the transport organized on their behalf. As Lieutenant Pritchard affirmed, 'Doctors and medical stores were present in abundance.' After Omdurman, he added, the wounded were spared the discomfort of being carried for long distances on stretchers; once their wounds were dressed, steamers conveyed them to the 'roomy, well-roofed, mud-brick hospitals' at Fort Atbara and Abadieh. 'When convalescent they travelled by alternate rail and steamer to Cairo.'[61]

Perhaps the greatest contrast, though, lay with the experiences of the troops on the Gordon relief expedition. Bennet Burleigh, the war correspondent of the *Daily Telegraph*, who had travelled and fought alongside the camel corps in 1884–85, regarded the Sudan Military Railway as 'astounding in conception'. By crossing 'the bare, unredeemed, and totally barren desert that lies betwixt Wady Halfa and Abu Hamed for 230 miles [370 km]', the railway, in his opinion, had 'entirely changed the whole situation in the Soudan', and 'brought the reoccupation of Khartoum within easy and measurable distance'.[62]

A decisive victory followed at Omdurman (2 September 1898),[63] and assessments of the campaign proliferated, with all manner of tributes paid to the sirdar's tactics, his use of overwhelming fire-power, the courage of the enemy, the changing of fronts by Hector Macdonald's Sudanese brigade as they faced charges from two different fronts, and the bravery and discipline, if not the tactical skill, of the Lancers' charge.[64] The more perceptive observers, though, understood that this remarkable victory in which 10,200 Mahdists had been killed, and possibly another 16,000 wounded in a morning, derived from the transformative effect of the Sudan Military Railway. On his return to London, the battle-scarred veteran of the Black Watch Colonel Andrew G. Wauchope, who had assumed command of the 1st British Brigade when Gatacre became the divisional commander, declared that 'it would have been quite impossible to have carried through the work so successfully if it had not been for the extraordinary foresight of the Sirdar in suggesting the construction of a railway across the desert from Wadi Halfa to the Atbara River'.[65] Hunter, who had described the 'Railway from Halfa to Abu Hamed' as 'a monument to the skill & resources of the Sirdar', and had predicted that 'The Railway is the all important factor of this expedition,'[66] now explained that the ability to concentrate fire-power on land and the river had bewildered the enemy. After the surprise at Firket, they had then fought 'behind rocks and in houses' at Hafir and Dongola, where the 'gunboats played Hell' with them; at Abu Hamed they had fought from houses but were 'attacked and defeated'; at the battle of Atbara (8 April 1898), Mahmud had reserved his fire (and hence inflicted 'shocking wounds and large numbers of amputations and deaths') before being overwhelmed; and finally at Omdurman, instead of attacking at night, the khalifa had fired first and launched a massive charge across an open plain to a catastrophic defeat.[67]

Churchill expressed the impact in more florid prose: the railway had enabled the sirdar to bring a considerable army into the heart of the Sudan and supply it 'with abundant food and ammunition, but with all the varied paraphernalia of scientific war'. He was also able to support its action on land 'by a powerful flotilla of gunboats, which could dominate the river and command the banks'.[68] That paraphernalia extended beyond a vast capacity for long-range fire-power to encompass searchlights on the gunboats, which protected the perimeter of Kitchener's defences through the night of 1 September, telegraphic communications from Cairo to the forward base at Fort Atbara and an X-ray apparatus to assist in treating the wounded at Abadieh.[69]

As an engineer, Pritchard was possibly somewhat biased, but he was convinced that everyone who took part in the campaign admired 'the Sirdar's genius and perseverance in having thought out and made such a perfect line of communications through such difficult and inhospitable country'. He reckoned, too, that the success and smooth working of the operations derived from 'the speed and thoroughness with which Lieutenant Girouard carried out the laying of the railway', and from the fact that so many men in responsible positions 'had spent several years on the frontier', acquiring a thorough knowledge of the country and its people.[70] If Kitchener's great triumph would be consummated by his diplomatic coup at Fashoda (19 September 1898), where he adroitly prevailed over Commandant Jean-Baptiste Marchand,[71] he returned to London to receive a hero's welcome, a peerage and the Freedom of the City of London. It was the legacy of this campaign, probably more than his subsequent services in South Africa and India, that established his reputation with the British public. The reassurance that he embodied when appointed as secretary of state for war at the outbreak of the Great War was an enduring reminder of this remarkable campaign and its iconic symbol, the Sudan Military Railway.

Notes

1 On the preference for fighting on the plains of Karari, see I. H. Zulfo, *Karari: The Sudanese Account of the Battle of Omdurman*, trans. P. Clark (London: Frederick Warne, 1980), pp. 120–1, 137–40; on the campaign as a whole, see H. Keown-Boyd, *A Good Dusting: A Centenary Review of the Sudan Campaigns 1883–1899* (London: Leo Cooper, 1986) and E. M. Spiers (ed.), *Sudan: The Reconquest Reappraised* (London: Frank Cass, 1998); and a study of Kitchener, J. Pollock, *Kitchener: The Road to Omdurman* (London: Constable, 1998).

2 G. W. Steevens, *With Kitchener to Khartum* (Edinburgh: William Blackwood & Son, 1898), p. 22.

3 I. F. W. Beckett, 'Kitchener and the Politics of Command' in Spiers (ed.), *Sudan: The Reconquest Reappraised*, pp. 35–53.

4 D. Johnson, 'The Death of Gordon: A Victorian Myth', *Journal of Imperial and Commonwealth History*, vol. 10, no. 3 (1982), pp. 285–310.

5 F. R. Wingate, *Mahdism and the Egyptian Sudan*, 2nd edition (London: Frank Cass, 1968), pp. 466–7, 469, 478, 491; Col. Sir R. Slatin Pasha, *Fire and Sword in the Sudan: A Personal Narrative of Fighting and Serving the Dervishes 1879–1895* (London: E. Arnold, 1896), p. 406.

6 Hatfield House Archives (HHA), 3rd Marquess of Salisbury MSS, vol. A113, no. 8, Lord Salisbury to Lord Cromer, 13 March 1896; see also E. M. Spiers, 'Introduction: The Reconquest Reappraised' and D. Steele, 'Lord Salisbury, the "False Religion" of Islam, and the Reconquest of the Sudan' in Spiers (ed.), *Sudan: The Reconquest Reappraised*, pp. 1–10 and 11–34.

7 HHA, 3rd Marquess' MSS, vol. A113, nos. 8 and 12, Salisbury to Cromer, 13 March and 1 April 1896.

8 Ibid. vol. A109, no. 24, Lord Cromer to Lord Salisbury, 15 March 1896; TNA, FO 78/4893, Lord Cromer to Lord Lansdowne, 22 April 1896; BL, MS 5/49, Lansdowne MSS, Lord Lansdowne to Lord Salisbury, 17 March 1896.

9 HHA, 3rd Marquess' MSS, vol. A109, nos. 24 and 48, Cromer to Salisbury, 15 March 1896, and Cromer to Salisbury, 7 May 1896.

10 Sandes, *Royal Engineers in Egypt and Sudan*, pp. 151, 161; Keown-Boyd, *A Good Dusting*, ch. 16; M. W. Daly, *The Sirdar: Sir Reginald Wingate and the British Empire in the Middle East* (Philadelphia: American Philosophical Society, 1997), p. 85.

11 'An Officer' (Lieutenant H. L. Pritchard, RE), *Sudan Campaign 1896–1899* (London: Chapman & Hall, 1899), pp. 18–19: Keown-Boyd, *A Good Dusting*, pp. 151–2; 'Colonel Edward Percy Cranwill Girouard', *RE Journal*, vol. 47, no. 2 (1933), pp. 323–43.
12 Earl of Cromer, *Modern Egypt*, vol. 2, p. 88.
13 Ibid. vol. 2, p. 89.
14 Sandes, *Royal Engineers in Egypt and Sudan*, pp. 155 and 160.
15 Ibid. p. 155; 'An Officer', *Sudan Campaign*, p. 19.
16 HHA, 3rd Marquess' MSS, vol. A109, no. 61, Cromer to Salisbury, 13 June 1896.
17 Ibid. no. 65, Maj. F. R. Wingate to Cromer, 10 June 1896.
18 Parl. Deb., Fourth Series, vol. 45, col. 1440 (5 February 1897); Sandes, *Royal Engineers in Egypt and Sudan*, p. 166 n. 2.
19 NAM, Acc. 1986-05-26-17, Burn-Murdoch MSS, TS diary, 13 August 1896, fo.120.
20 'An Officer', *Sudan Campaign*, p. 49; Keown-Boyd, *A Good Dusting*, p. 152. Several sources claim that he required over 5,000 men to repair the line: P. Magnus, *Kitchener: Portrait of an Imperialist* (London: John Murray, 1958), p. 98; A. B. Theobald, *The Mahdiya: A History of the Anglo-Egyptian Sudan, 1881–1899* (London: Longmans Green, 1951), p. 204.
21 E. F. Knight, *Letters from the Sudan* (London: Macmillan & Co., 1897), p. 223.
22 Sandes, *Royal Engineers in Egypt and Sudan*, pp. 167–8.
23 Sudan Archive Durham University (SAD), SAD, 304/2/1–34, Maj. J. J. B. Farley, 'Some Recollections of the Dongola Expedition', pp. 23–5.
24 Ibid. p. 26.
25 SAD, Hunter MSS, SAD, 964/2/28–31, Col. A. Hunter to Capt. J. R. Beech, 23 July 1896; see also A. Hunter, *Kitchener's Sword Arm: The Life and Campaigns of General Sir Archibald Hunter* (Staplehurst: Spellmount, 1996).
26 Sir G. Arthur, *Life of Lord Kitchener*, 3 vols. (London: Macmillan & Co., 1920), vol. 1, p. 205; on the state of this railway, see D. A. Welsby, *Sudan's First Railway: The Gordon Relief Expedition and the Dongola Campaign*, Sudan Archaeological Research Society Publication 19 (London: Sudan Archaeological Research Society, 2011).
27 HHA, 3rd Marquess' MSS, vol. A109, no. 98, Cromer to Salisbury, 14 October 1896.
28 Ibid. no. 102, Cromer to Salisbury, 30 October 1896.
29 Ibid. vol. A113, no. 34, Salisbury to Cromer, 27 November 1896: see also Magnus, *Kitchener*, p. 102.
30 Kitchener may have decided on the route as early as April 1896: Sandes, *Royal Engineers in Egypt and Sudan*, p. 192 n. 1; see also Magnus, *Kitchener*, pp. 104–5, and W. S. Churchill, *The River War: The Sudan, 1898* (London: Eyre & Spottiswoode, 1899, reprinted London: Sceptic, 1987), p. 173.
31 NAM, Acc. no. 1968-07-234, Kitchener–Wood MSS, Kitchener to Sir E. Wood, 5 June 1897; see also Churchill, *River War*, p. 174.
32 H. S. L. Alford and W. Dennistoun Sword, *The Egyptian Soudan: Its Loss and Recovery* (London, 1898, reprinted London: Naval & Military Press, 1992), p. 163.
33 HHA, 3rd Marquess' MSS, vol. A110, no. 41, Kitchener, 'The Abu Hamed Railway', 6 June 1897; see also Sandes, *Royal Engineers in Egypt and Sudan*, p. 227.
34 Churchill, *River War*, p. 175.
35 Ibid. p. 177; Sandes, *Royal Engineers in Egypt and Sudan*, pp. 227–35.
36 NAM, Acc. no. 1968-07-234, Kitchener–Wood MSS, Kitchener to Wood, 3 July 1897.
37 TNA, PRO 30/40/2, fos. 556–7, Ardagh MSS, 'Extracts from Major à Court's No. 8', 12 November 1897; Hunter, *Kitchener's Sword-Arm*, pp. 68–70; Theobald, *The Mahdiya*, pp. 216–17; Sandes, *Royal Engineers in Egypt and Sudan*, p. 243.
38 NAM, Acc. no. 1968-07-234, Kitchener–Wood MSS, Kitchener to Wood, 30 September 1897; see also Keown-Boyd, *A Good Dusting*, p. 186.
39 HHA, 3rd Marquess' MSS, vol. A110, nos. 63 and 64, Cromer to Salisbury, 22 October 1897, and Kitchener to Cromer, 18 October 1897.
40 NAM, Acc. no. 1968-07-234, Kitchener–Wood MSS, Kitchener to Wood, 19 November 1897; see also TNA, FO 78/5049, 'Intelligence Report, Egypt, No. 56, 6th October

to 12th November 1897', p. 7, and HHA, 3rd Marquess' MSS, vol. A110, no. 66, Cromer to Salisbury, 22 October 1897.

41 E. M. Spiers, *Wars of Intervention: A Case Study – The Reconquest of the Sudan 1896–99*, The Occasional, 32 (Camberley: Strategic & Combat Studies Institute, 1998), pp. 16–17.

42 'An Officer', *Sudan Campaign*, p. 40. For slightly different mileage, see Churchill, *River War*, p. 182.

43 TNA, FO 78/4895, Cromer, telegram, enclosing Sirdar's reply, 25 December 1897.

44 HHA, 3rd Marquess' MSS, vol. A110, no. 88, Cromer to Salisbury, 25 December 1897.

45 SAD, 102/1/69, Wingate diary, 31 December 1897.

46 TNA, FO 78/5049, nos. 1 and 10, Cromer to Salisbury, 1 January 1898 and 6 January 1898.

47 Churchill, *River War*, p. 183; see also Sandes, *Royal Engineers in Egypt and Sudan*, p. 238.

48 'An Officer', *Sudan Campaign*, p. 169; Sandes, *Royal Engineers in Egypt and Sudan*, p. 238.

49 NAM, Acc. no. 1968-07-34, Kitchener–Wood MSS, Kitchener to Wood, 17 July 1898.

50 Theobald, *The Mahdiya*, p. 222.

51 Museum of Lincolnshire Life (MLL), LR 770, Capt. S. Fitzgibbon Cox, diary, 7 January 1898.

52 SAD, 643/1/7, Longe MSS, 2nd Lt. H. P. Creagh-Osborne, diary, 19 January 1898; 'A Drummer of the "Sixth" on the Soudan Campaign', *Warwick and Warwickshire Advertiser & Leamington Gazette*, 30 April 1898, p. 6.

53 'A Cumbrian in the Soudan', *Newcastle Daily Chronicle*, 10 October 1898, p. 3; see also 'A Horncastrian at the Battle of Atbara', *Horncastle News and South Lindsey Advertiser*, 28 May 1898, p. 5, 'The Camerons at Omdurman', *Inverness Courier*, 7 October 1898, p. 5, and 'Letter from a Nairn Man at Atbara', *Nairnshire Telegraph*, 25 May 1898, p. 3.

54 Guards Museum, Jeffreys diary, 31 July and 1 August 1898. The Grenadier Guards seem to have endured a particularly miserable journey in Lower Egypt: 'a very tiring ride', wrote Lance-Sergeant George Shirley, 'In the Soudan Campaign', *Hampshire Observer*, 8 October 1898, p. 3.

55 Queen Mary, University of London Archives, Lyttelton Family Papers, NG/KL/498a, Brig. Neville G. Lyttelton to Katherine Lyttelton, 5 August 1898.

56 NAM, Acc. no. 1978-04-53, Capt. D. W. Churcher, TS diary, 7 August 1898.

57 H. F., 'Our Baptism', *Vedette*, no. 109 (1898), pp. 2–22, at p. 3.

58 NAM, Acc. no. 1974-04-36-3, Lt. R. F. Meiklejohn, 'The Nile Campaign', p. 4, and Acc. 1978-04-53, Churcher, TS diary, 7 August 1898 (4.30 p.m.).

59 'With the Army Medical Corps', *Sheffield Daily Telegraph*, 4 October 1898, p. 6.

60 NAM, Acc. no. 1974-04-36-3, Meiklejohn, TS diary, p. 11; see also MLL, Fitzgibbon Cox, diary, 27 February, 1, 3 and 4 March 1898, and J. W. Stewart, 'A Subaltern in the Sudan, 1898', *The Stewarts*, vol. 17, no. 4 (1987), pp. 223–8, at p. 225.

61 'An Officer', *Sudan Campaign*, p. 228.

62 B. Burleigh, *Sirdar and Khalifa or the Reconquest of the Soudan 1898* (London: Chapman & Hall, 1898), pp. 96–7.

63 Although the battle of Omdurman broke the military power of the khalifa, he fled the field and remained a rallying point for the disaffected until he was killed at the battle of Um Dibaykarat (24 November 1899).

64 E. M. Spiers, 'Campaigning under Kitchener' in Spiers (ed.), *Sudan: The Reconquest Reappraised*, pp. 54–81, at pp. 69–73; see also Keown-Boyd, *A Good Dusting*, pp. 227–36, and P. Ziegler, *Omdurman* (London: Collins, 1973).

65 'Colonel Wauchope on the Soudan Campaign', *Edinburgh Evening News*, 1 October 1898, p. 3.

66 Liddell Hart Centre for Military Archives, King's College London, Maurice MSS, 2/1/4, Col. A. Hunter to Maj.-Gen. Sir J. F. Maurice, 15 February 1898.

67 SAD, Hunter MSS, SAD, 964/4/64–73, Hunter, letter from Khartoum, 14 October 1898.
68 Churchill, *River War*, p. 183.
69 Spiers, *Wars of Intervention*, pp. 30–33, 48, and R. F. Mould, *A Century of X-Rays and Radioactivity in Medicine: With Emphasis on Photographic Records of the Early Years* (Bristol: Institute of Physics, 1993), pp. 56–7.
70 'An Officer', *Sudan Campaign*, pp. 174 and 261.
71 J. F. V. Keiger, 'Omdurman, Fashoda and Franco-British Relations' in Spiers (ed.), *Sudan: The Reconquest Reappraised*, pp. 163–76.

7

Railways on the veld: the South African War, 1899–1902

Just as the South African War (1899–1902) tested many facets of the late Victorian army – its command, staff, tactics and means of expansion, and its operational and supply capacity – so it proved the ultimate test of the army's ability to use, manage and exploit railways in wartime. Prior to the outbreak of war, there were 8,083 km of railroad in southern Africa (see Map 1 at the front of this book), most of it narrow-gauge (3 feet 6 inches, or 107 cm) and single-track. Approximately 7,448 km were in direct intercommunication with one another, of which 5,258 km

Figure 9 'Outside Ladysmith: Night-Signalling with the Search Light from an Armoured Train', *Illustrated London News*, vol. 116, no. 3171 (27 January 1900), p. 120

were in the British colonies of Natal and Cape Colony, 544 km in the Orange Free State, 918 km in the Transvaal and 88.5 km in Portuguese territory. The discoveries of diamonds in Griqualand West in 1867 and gold in the Witwatersrand in 1886 had stimulated much of the railway building, with the railroad reaching the 'Diamond City' of Kimberley on 28 November 1885. The steep gradient from the Cape into the interior reduced the speed of the journeys, and meant that it took thirty-three hours to reach the junction of De Aar, some 650 km north-east of Cape Town. From Natal the main line ran from Durban via Pietermaritzburg to Ladysmith, and thence north-west towards the Natal–Transvaal border, crossing the Biggarsberg mountain range, while a branch line from Ladysmith went west into the Orange Free State.[1]

British reliance upon these lines of railway became a source of controversy both during the war and in subsequent historiography. 'The feeding, reinforcing, and maintenance of a large force by means of one single artery or line of railway is a practical impossibility', observed the *Newcastle Daily Chronicle* in May 1900. Even if all went well at first, and units passed forward to the front, 'the number of mouths to feed causes a daily increasing strain on supply trains and railway staff'. When an accident occurred, as one did in March 1900, blocking the entire Cape Town–De Aar line, it took place 'just when large numbers of remounts and fresh mounted troops were urgently required by General French . . .'.[2] Operationally, too, the single-track railway could encumber a commander like Paul Stanford Methuen, Lord Methuen, in his advance to relieve the besieged citizens of Kimberley, acting 'like a drag-rope upon his movements'.[3] Douglas Porch even asserted that the 'British advances along rail lines in the early months of the Boer War were so predictable that the Boers simply had to fortify obvious choke points and wait for the British to attack.'[4] The Boer commandos, argued Ian van der Waag, 'almost invariably mounted, broke British railway lines more than 250 times in 12 months', before disdaining any further commentary on the significance of the railroads.[5] These cryptic comments contrast with more informed commentaries, including a detailed history of the railway war, a major chapter in the sixth volume of *The Times History* and a more recent, concise account that provides insights into the Boer use of railways.[6]

This chapter will not simply review the ways in which railways were used in the war, but will also seek to explain the significance of the operational movements at the outset of hostilities, including the sieges of Mafeking (Mafikeng), Kimberley and Ladysmith and the attempts to relieve them. It will examine how travelling, protecting and working on the railways affected the wartime experiences of

British soldiers, the vast majority of whom had never been to South Africa before. It will comment upon the management of railway assets after Lord Roberts assumed overall command, following the early reverses at Stormberg, Magersfontein and Colenso, the 'Black Week' of 11 to 15 December 1899, and then broke free from the railway in his push towards Bloemfontein, the capital of the Orange Free State. It will consider, too, how the use and conflict over railway assets evolved, as British forces drove through the Boer republics and then faced sustained attacks on the railroads once the guerrilla war began. Underpinning these assessments will be reflections upon how the British refined their use of armoured, transport and hospital trains, and maintained a massive logistics network largely through hostile and often waterless country. Finally, it will reflect upon the changing nature of the war, as the British moved Boer families and many blacks to encampments supplied from the railways, and employed a huge network of blockhouses to protect the railway lines and assist in the 'drives' that characterized the concluding months of the war.

Railways during the Boer offensive

The South African War did not just 'break out';[7] after a protracted period of negotiations, the Boer republics delivered an ultimatum to the British government on 9 October 1899, demanding *inter alia* the removal of British forces from the republican frontiers. When this was rejected, the Boer republics launched their invasions of Cape Colony and Natal on 11 October 1899. Within two days Boer forces under General Piet Cronjé besieged the small frontier town of Mafeking, seeking to ensure that the British did not encourage the black communities along the Transvaal border to attack the commandos and their civilians. The siege followed the opening engagement of the war, when Boers under General Koos de la Rey captured an armoured train at Kraaipan on 12 October.[8] Following this embarrassing start to hostilities, the small British garrison in the 'Diamond City' of Kimberley found itself besieged when Free Staters cut the town's telegraph lines on the evening of 14 October. This began the 124-day investment of a town famous for its mineral wealth, its most illustrious citizen, Cecil J. Rhodes, and its railway connection over 1,041 km to Cape Town. Although the town had prepared elaborate defences, large numbers of refugees fleeing from the Boer republics became trapped in Kimberley, and so the local railway manager sought to boost the store of foodstuffs by retrieving a large consignment of flour from the nearby Modder River station. As one of the refugees, J. Fred Byrne, a former rugby football player for Moseley, England and the British Lions, recalled, 'Hundreds of sacks of flour were taken into Kimberley that Saturday afternoon, a few hours afterwards the Boers had blown up the railway culverts and isolated the town . . .'.[9]

During this opening offensive the Boers, though renowned for their equine mobility, exploited the railway network in a comprehensive fashion. Prior to the war on 13 September 1899, the Transvaal government had taken over the railways previously owned by the *Nederlandsche Zuid-Afrikaansche Spoorweg-Maatschappij*, and, once the war began, both republics used railways to transport burghers, their supplies and heavy ordnance into Natal and to reinforce commandos on the southern front. In fact, the Boers moved much of their artillery, including two of their four 15-mm Creusot guns, later nicknamed 'Long Toms', to their different borders before the onset of hostilities. They were able thereby to deploy 'Wonderboom Tom' just inside the Natalian border to protect the route into the Transvaal, while the burghers, with another Long Tom and other artillery, invaded Natal. Effectively the Transvaalers exploited Natal's rail route from Colenso, Ladysmith and Newcastle, which crossed the Transvaal border at Volkrust and proceeded to either Pretoria or Johannesburg via Elandsfontein.[10] Further south Free Staters pressed through the Natalian border, partly along the branch line that came from Harrismith to Ladysmith.

At least at its outset the Natal campaign had to be conducted near the railway lines. This is where the principal Natalian settlements were located, where Ladysmith served as the major railway junction north of the Tugela (Thukela) River, and where Lieutenant-General Sir George White, VC, had concentrated his forces, including recently arrived reinforcements from India. After a pyrrhic victory at Talana Hill (20 October 1899), British forces found themselves outnumbered, and outgunned, by the invaders. White sent another force north from Ladysmith to confront the Boers under General Johannes Kock, who were pressing southwards. Lieutenant Walter W. MacGregor (2nd Battalion, Gordon Highlanders) described how his men 'marched to the station, straight into the train, went for almost three-quarters of an hour, detrained near a small station called Modder Spruit, and immediately extended for attack on Elandslaagte'.[11]

After another costly victory at Elandslaagte (21 October 1899), British forces used the railway to evacuate their wounded relatively quickly from the field of battle. An officer of the 1st Battalion, Manchester Regiment, described how he 'was put on a stretcher and carried about three miles [4.8 km] to the railway, where our regiment was drawn up to cover the entrainment of the wounded'.[12] The platform and station yard at Ladysmith, though, struggled to accommodate these casualties: Nurse Kate Driver recollected that the wounded 'were crowded with rows and rows of stretchers, d'Hoolies, as we called them in Natal. As we threaded our way through them we heard moans of agony coming from under the green covers.'[13] Finally, after the British defeats at Lombard's Kop and Nicholson's Nek, 'Mournful Monday' (30 October 1899), the siege of Ladysmith began three days later, allowing Major-General John D. P. French and Colonel Douglas Haig to escape on the last train from the town on 2 November. The Boers now found another use for railways, namely transporting British prisoners back from northern Natal and on to Pretoria. Meeting Boers probably for the first time, one gunner recalled how 'Our captors have been most kind to us . . . At Volkrust a commissariat officer (Boer) got two bottles of whisky – all that was obtainable – and gave it to us. The people at the stations were most curious to see us.'[14]

Railways, as ever, served as an important point of social interaction. As many *Uitlanders*, the white miners and their families or dependants, fled from the Transvaal, they had fraught experiences on the railway. One refugee described the week-long journey from Johannesburg to Cape Town in a train composed of forty coal trucks and six carriages: 'At night passengers were required to sleep on the veldt. Whenever a stoppage was made, the occupants of the train were made the butt of Boer insults . . . and found the muzzle of rifles placed under their noses by cowards who called them "venom roineks" to their faces.'[15] The refugee trains travelling from the Transvaal appalled a telegraphist working at De Aar junction:

> I have just seen a sight I hope I may never see its like again – forty open cattle trucks densely packed with men, women, and children, refugees travelling to Cape Town, and all craving for food. It was pitiful to see them in the blazing hot sun, with a fearful wind laden with fine dust . . . They ate like ravenous wolves. It was heartrending to see them.[16]

Nor were such scenes confined to the days immediately before and after the outbreak of war: after Bloemfontein fell to the British on 13 March 1900, Fred Wall, part of the local British community, witnessed 'refugees from Pretoria, Johannesburg, and other towns north, passed through in trucks packed and a lot of them starving – blacks and whites all mixed up anyhow. Really it was terrible to see.'[17] Meeting a defeated and chastened Boer prisoner on the railway was a different proposition; a soldier of the 2nd Battalion, Duke of Cornwall's Light Infantry, accompanied a party of twenty-eight Boer prisoners on the railway from Orange River to Cape Town, and found them 'a decent lot of men, and, in fact, could not do too much for the six of us forming their escort . . . They must have spent, at least, £10 on us. . . .'[18]

Even more positively, railways provided splendid opportunities for English-speaking communities to display their support for the troops and for the war. In Natal, where the citizens of Durban had already greeted British soldiers on their arrival from India, demonstrating a peculiarly fond memory of the Gordon Highlanders, who had served in the previous Anglo-Transvaal War,[19] there was even more enthusiasm after the siege of Ladysmith began. As groups of Boers probed south of the Tugela River, cutting the railway and seeking defensible positions, incoming British soldiers, who were part of the relief force, wrote of their ecstatic receptions in the colony. A soldier of the 2nd Battalion, Somerset Light Infantry, described how the Natalians rallied as they travelled up to the Mooi River: 'We had a warm reception coming up the line. We got everything given us at different stations – lemonade, fruit, and tobacco. The people cheered us at every station.'[20]

Relief operations in Natal and Cape Colony

Sir Redvers Buller, VC, the original commander-in-chief of the Army Corps, South Africa, who arrived in Cape Town just after 'Mournful Monday, changed his strategy and sent large forces by railway up to relieve Ladysmith and Kimberley. Splitting his corps proved controversial, especially when it ended in the defeats of 'Black Week',[21] but the logic of using the railways meant that he could move large forces into the interior, accelerate the prospect of a fast relief of the embarrassing sieges, and curb any prospect of the Boers rising in revolt within Cape Colony. He despatched a division of about 10,000 men, including elite Guards and Highland Brigades, under Lord Paul S. Methuen towards Kimberley, and another division under Sir William Gatacre to the railway junction at Stormberg to secure the Cape Midlands district from Boer raids and local rebellions by Cape Boers. He then accompanied a third division to Natal, advancing up the railway line towards Colenso station, where the Boers had constructed their principal defensive positions on the north bank of the Tugela River. Although the Boers anticipated the rail-based lines of advance, in both Natal and Cape Colony, British forces had a preponderance of manpower and guns, if not mounted forces, in the ensuing engagements. The heavy casualties incurred in the three victories on the western front reflected failures of reconnaissance (Belmont), an inability to manoeuvre or pursue a retreating enemy (Graspan) and more failings in reconnaissance, a faulty map and intelligence analysis (Modder River). Failures of reconnaissance and intelligence, compounded by lapses in command, staff work and tactical blunders, recurred in all three defeats of 'Black Week'.[22] Railways may have enabled the relief forces to engage the enemy, and boost the morale of the besieged by establishing night-time communications from search-lights on the armoured trains (see Figure 9) but they neither determined how commanders engaged the enemy nor accounted for all their failings in battle.

While the British government responded to the 'Black Week' defeats of Stormberg, Magersfontein and Colenso by appointing a new commanding officer, Lord Roberts, and by sending out further reinforcements, it accepted the services of volunteers at home and from Canada, Australia and New Zealand. Girouard, as the incoming director of railways, seized this opportunity. Only thirty-two years of age, and holding the local rank of lieutenant-colonel, he had arrived in South Africa on 23 October to confront a host of challenges in managing the massive railway network (see Figure 10). Initially he had the support of only two railway companies, RE (the 8th and 10th), which had been brought up to war strength (ten officers and 300 other ranks), but, by November, he had secured the re-assignment of another two RE fortress companies (the 31st and 42nd) for railway work. In addition he formed a staff of enterprising officers, who possessed experience of railway work in Canada, India and the Sudan, and had the co-operation

Figure 10 Lieutenant-Colonel Edward 'Percy' Girouard, RE

and professional skill of the staff of the Cape Government Railways (CGR) and the Natal Government Railways. Accordingly, he exploited the period of inaction that followed the 'check at all points towards the middle of December' as

> of the greatest advantage to the railway departments. It enabled supplies and railway material to be forwarded to the front; it gave time for the elaboration of the new organization, and for the formation of the Field Railway Sections and Railway Pioneer Regiment . . .[23]

In fact, Girouard benefited even more from the defensive strategy adopted by the enemy. Although the Boers had invaded Cape Colony and Natal, they were not bent upon conquest and annexation. Hoping to repeat the triumphs of the Anglo-Transvaal War, they had preferred to mount sieges, defeat the British outside their own territories and seek a negotiated settlement. Inhibited by their deep religious convictions, they did not want to strike first and assume a full-scale military offensive. They relied, too, upon a citizen army, which lacked the discipline of their British counterparts, and so had to avoid unnecessary casualties. Marthinus Steyn, the president of the Orange Free State, refused to attack the railway lines of Cape Colony, prompting Colonel John F. Y. Blake, an Irish-American who fought with the Boers, to despair of the consequences: 'So De Aar Junction and all the railway lines were allowed to remain in good order for the use of Lord Roberts and his army.'[24]

Profiting from the enemy's instinctive caution, Girouard claimed that the railway staff under his orders had two overriding aims:

> (a) To keep the military commanders fully informed of the capacity and possibilities of the railway and to convey their orders and requests to the civil railway staff.
>
> (b) To protect the civil railway administration from interference by military commanders and commandants of posts; in fact, to act as intermediaries between the Army and the civil railway officials.[25]

In order to fulfil these aims, Girouard's staff had to retain authority over the railway. No other officer should be able to give orders to railway staff officers or other railway officials, unless fighting was taking place in the vicinity. Within the remit of the Cape railways, an assistant director of railways served as the counterpart to the general traffic manager, CGR, four deputy assistant directors of railways served as the counterparts to the traffic managers of the four railway sections of the CGR, and railway staff officers served at all the important stations. The intermediary controlling staff would in due course establish a uniformity of administration across the railway network (other than in Natal, where the military demands, the railway mileage and the enemy threats after Buller relieved Ladysmith were proportionately much less than elsewhere). The system also enabled correct accounts to be kept between the army and civil companies, based on a uniform rate per truck per mile. Moreover, during December 1900, Girouard accepted the offer of Sir Alfred Milner, the high commissioner, to employ 1,000 miners and artisans who had fled from the goldfields and were out of work in Cape Town. They comprised the Railway Pioneer Regiment, and assisted in making permanent and semi-permanent repairs to the rail lines, following up temporary repairs made by the Royal Engineers. After Kimberley was relieved (15 February 1900), Lieutenant-Colonel R. G. Scott, VC, raised Scott's Railway Guards, a corps of about 500 men principally from the De Beers Mines, to guard the Orange River–Kimberley line while the Pioneers sought to protect the Central Railway.[26]

In mid-January 1900 Lord Roberts and his chief of staff, Lord Kitchener, arrived in Cape Town, bringing vast numbers of reinforcements and great quantities of military stores. They began raising new bodies of mounted infantry – Roberts's Horse and Kitchener's Horse – and all these troops, horses, guns and equipment had to be moved north, concentrating initially in the neighbourhood of Naauwpoort and Rosmead. When Roberts ordered his famous change of front, railing all these forces westwards to detrain between the Orange and Modder Rivers, further reinforcements were still pouring in from Cape Town and Port Elizabeth and increasing the strain on facilities and railway management at the junction of De Aar. The works departments of both the Midland and Western Railway Sections had to lay an additional 16 km of sidings at Orange River, Graspan, Enslin, Honey Nest Kloof and Modder River to accommodate the troop movements. Thereafter, as Girouard explained, the railing of forces westwards required the 'strictest secrecy', so ensuring that units received only minimal information at the last minute, and even then only in code. The whole process involved 152 trains, comprising 4,787 trucks, passing northwards through De Aar between 26 January and 10 February, and required detailed planning. It involved the superintendence of railway staff officers at various detraining stations by officers of the Royal Engineers and Cape government officials; the entrainment of a large body of cavalry at the Rensburg siding; and traffic management at the key stations of Naauwpoort, De Aar and Orange River. Ultimately Girouard reckoned that during the period from 26 January to 12 February 1900 his team detrained 30,000 soldiers, with horses, mules, oxen, guns and transport between Orange River and Modder River: 'This was the largest troop move carried out by the railways during the campaign, and the greatest credit is due to the Cape Government Railways for the manner in which it was worked.'[27]

Railways proved indispensable at this stage in the campaign. They enabled the British army to deploy and support unprecedented numbers of men and livestock in the field in a remarkably short period of time. These demands differed sharply between the colonies of Natal and Cape Colony; in the former colony, British relief forces simply moved from Durban to the railhead, a distance of some 258 km, but then rarely travelled by train (apart from General Hunter's division, which later travelled from the railhead to the sea). Nor was the scale of operations as demanding as in Cape Colony. In Natal British forces never exceeded 40,000 men, and the Natal railway rarely encountered enemy disruption after the relief of Ladysmith. Although numerous repairs had to be made to the line and tunnels north of Ladysmith, the Natalian line (then 595 km in length) experienced only seven breaks up to May 1901. As a consequence, the Natal railway 'fed General Buller's Army with the greatest ease, and when the Transvaal was annexed it became one of the main sources of supply to the whole Army'.[28]

In Cape Colony De Aar, as Blake anticipated, became a critical hub. Based at De Aar, the assistant director of railways, Major V. Murray, RE, worked with Mr Clark, the traffic manager of the Western Section of the CGR, to control all the movements during the change of fronts. The station had emerged from humble beginnings. It was originally a mere railway camp with a parallelogram of huts enclosing a post office, a hotel and an institute 'where the drinking of strong liquors is forbidden . . . It has been called the Clapham Junction of South Africa, as it was from this point the railway traffic debouched for Kimberley, for the Free State, and for Pretoria, as well as for Bulawayo and Natal itself.'[29] Protected to the north by an open plain to the southern bank of the Orange River, the plain connected De Aar with Naauwpoort and Colesberg. Only one infantry battalion protected the line initially and guarded the stores. 'The worst of this vile place', wrote a British officer, 'is that there is not a loyal native within twenty miles [32 km] of us, and they are only waiting for an opportunity to rise.'[30] This was a slight exaggeration, as the King's Own Yorkshire Light Infantry under Major Henry Earle were constructing 'some formidable redoubts and earthworks' in October 1899 to protect the base. As the earthworks brought the reassurance of a permanent British presence, it may have softened the views of local inhabitants: at least three soldiers were found 'helpless and incapable' at the earthworks after receiving liquor from 'native women'.[31] By December 1899, as 'punishing' sand storms swept through the camp, Private Charles Jones (Gloucestershire

Mounted Infantry) found that De Aar had become 'a large military camp with plenty of dust and not much water'.[32] In the following January, it was still 'a bit of red hot sand' where soldiers were 'planted in ignominious safety . . . swearing from morning till night, at a kind of work which any self-respecting Punjabi coolie would consider undignified'.[33] However demeaning, the work transformed De Aar into a secure and functioning railway junction that sustained British operations both within and from Cape Colony.

Just as the operations differed markedly between Natal and Cape Colony, so did the railway experiences of the incoming forces. In the Cape, Sergeant W. Hart-McHarg (Royal Canadian Regiment) recalled, the railway had

> an excellent road-bed, solid and smooth, being superior in this respect to the Canadian and American lines, and the bridges and culverts are all iron and stone; but it is all grades and curves, and a narrow gauge into the bargain. At one moment the train rushes along at sixty miles an hour, a minute or two later passengers can get out and walk . . .[34]

During the journey to De Aar, he added, the train stopped three times a day 'to make tea or coffee, rations of bread and meat also being served out. The people were quite enthusiastic all along the line . . . and welcomed us with cheers and waving of pocket handkerchiefs.'[35]

Conversely, the Natal authorities were notorious for their appalling treatment of British soldiers. Despite charging the army for third-class travel, they provided open coal trucks for the journey from Durban, often cramming forty men with all their arms and kit into each truck. If it rained for the entire journey, as endured by the 2nd Gloucesters, and later the 2nd Battalion, Royal Fusiliers, it was a thoroughly miserable experience.[36] A Swindonian, probably from a railway background, was amazed at 'how much solid discomfort can be crowded into a journey of less than 200 miles [322 km]'. Describing a return journey from Ladysmith to Durban, he dwelt first upon the 'extraordinary uncovered vehicles that a casual observer would call trucks', masquerading as carriages. Then

> We are off: with a bumping and thumping that emphatically prove that vacuum brakes are conspicuous by their absence . . . Now the rain and wind is upon us, beating in our faces without mercy, and what is worse beating down the steam, smoke, and dust as it leaves the engine's chimney, till every soldier whose face is toward the engine begins to assume a grimy streaky appearance . . .
>
> The country is blotted out. We cannot see, for rain and dust prevents one from opening his eyes.
>
> Thank Heaven, the rain is over. But what has happened to the singers?

> Perhaps the novelty most noticeable in travelling on this line is the flippant manner in which the engineers have treated curves that would make the hair of their confreres of the G.W. Railway stand on end!
>
> We race along the banks of some deep spruit [small stream] driving apparently straight into a high hill, then just as we expect a tunnel or something the engine doubles on its own tracks and comes back alongside us on the other side of the donga [dry watercourse].
>
> We reach the curve! The trucks tilt over to an alarming angle! It is going over! No. We resume the level on the back track! We heave a great sigh of relief, and laugh at our fears and one another, as we let go the seat we have been frantically clutching till we come to the next curve.
>
> So the journey continues, sometimes crawling up slopes as steep as 1 in 25, sometimes careering madly down the opposite slope, the engine whistling and shrieking like an animated demented creature. Up and down, and round about – a veritable switchback. And all the while the cramped positions one has to assume is telling its tale, in stiffening limbs; till when after a journey of apparently interminable length, we reach our destination at Durban, and detrain – as sorry a crowd as ever[y] wore Her Most Gracious Majesty's uniform.[37]

In both theatres relief forces assisted in repairing the rail track. In Natal, after the relief of Ladysmith (28 February 1900), Rifleman T. Plowman described how relief forces moved merely 32 km north of the town before they had to begin 'repairing the railway' (some thirty bridges and culverts to Glencoe), and later had to rebuild the damaged tunnel at Laing's Nek, which had had both ends blown in as well as about 45.7 m of lining at either end, consisting of big stone blocks.[38] Yet railway repairs were only one of several duties performed by the engineers, including 8th Company, RE, and a detachment of the Crewe reservists,[39] serving in Cape Colony. During the battle of Graspan (25 November 1899), they carried ammunition while under fire for three hours for the 4.7-inch Naval gun before forming a skirmishing line and supporting the infantry. At the ensuing battle of Modder River (28 November 1899), they supported the Argyll and Sutherland Highlanders and the Loyal North Lancashires after they managed to turn the Boer flank. 'The Crewe lads', wrote Sapper Champion, 'went through it like men. We had not much chance of firing, but we stopped the Boers from retiring that way.'[40] Subsequently, they assisted in repairing the damaged line and culverts, and in throwing railway and pontoon bridges for the troops across the Modder River.[41] Sergeant W. D. Western, RE, recalled how his company spent three days at the Orange River, repairing lines and culverts, before travelling up the line to the Modder River, where his men encamped: 'I am in charge of a party of R.E.'s [*sic*] cutting brushwood, and the sun is boiling. We have no water here.[42] The Boers broke the pipes and engines. The water comes up in tanks from Rensburg by train. We had water from the engine for tea last night.'[43] Constructing a railway bridge that could take trains across the Modder River required 'working from four in the morning till eight at night', wrote Sergeant J. Millard, RE: 'We have to work like niggers, and there are hundreds of Kaffirs with us. They, the Boers, blew two portions of the Modder River bridge down, and so we put a temporary one right across in eight days. . . .'[44]

This was doubly impressive since 'every foot of sleeper, every yard of timber, every nail, every screw even, has had to be brought with us from Capetown',[45] and the Boers had proved themselves highly proficient in the art of destruction. Trooper Claude Harrison, 12th Royal Lancers, observed that

> The Boers are a dead mark on blowing up railway bridges and telegraph wires. All along the rails there are small bridges and iron telegraph posts all blown to pieces. Our Engineers are having a terrible hard time of it, working night and day putting up wires and patching up bridges so as to be safe for traffic. There is a very big iron bridge over the river here, which the Boers have blown up at both ends. The Engineers have had to construct another line and bridge to get across.[46]

The repairs proved timely, as they not only enabled Lord Methuen to bring reinforcements forward but also facilitated the removal of hundreds of wounded soldiers after the disastrous defeat at Magersfontein (11 December 1899). The wife of a resident at De Aar saw trainload after trainload of sick and wounded reach the junction, where the hospital accommodated only 400 patients: 'there is no room in hospitals here', she feared, 'so they are going to take them on to Cape Town; fancy 500 miles [805 km] more in the train'.[47] More positively, the railway allowed Lord Roberts to assemble his vast army in the north of Cape Colony, despatch Major-General French with a mounted division of unprecedented size to relieve Kimberley, and then advance into the Orange Free State. He would defeat General Cronjé's army at Paardeberg (27 February 1900) before marching on to and capturing the Free State capital, Bloemfontein (13 May 1900).

The early use of trains to remove the wounded was just as conspicuous in Natal, where two hospital trains were fitted out at the beginning of the war. Another train, known as 'Princess Christian's Train', designed in England and erected in Durban, began running on 17 March 1900 (and was the first train to enter Ladysmith after the siege). Some hospital trains, usually with seven coaches, carried lying patients and each had a staff of two medical officers, two nursing sisters and twenty-two other ranks; others were made up of first-class corridor coaches for the less severe cases and convalescents, and some had a kitchen car attached; and then ambulance coaches gathered small groups of sick and wounded from various posts before being attached to passing trains for the offloading of passengers at the nearest hospital. The offloading was still a 'very distressing sight', as recalled by Sir William MacCormac, the consulting surgeon to the forces, but at least he described the sight at Chieveley station, only 6.4 km from the battlefield at Colenso.[48]

Just as distinctive were the operations involving armoured trains (see Figure 11) in both theatres. Used primarily as reconnaissance vehicles in this phase of the war, armoured trains proved highly vulnerable but none more so than the one commanded by Captain Aylmer Haldane from Estcourt to Chieveley on 15 November 1899, which carried Winston S. Churchill, the war correspondent of the *Morning Post*, some platelayers and 120 soldiers from the Royal Dublin Fusiliers and the Durban Light Infantry. The train comprised an ordinary truck in which six sailors manned a 7-pounder muzzle loading gun, an armoured car fitted with loopholes, the engine and tender, two more armoured cars and one more truck, loaded with tools and materials for repairing the line. Attacked when rounding a curve on the return journey from Chieveley, the train accelerated in trying to escape and

Figure 11 'A Sortie with the Armoured Train from Ladysmith', *Illustrated London News*, vol. 115, no. 3164 (9 December 1899), p. 832

ran into rocks placed on the line. This derailed the equipment truck and an armoured truck, and left the second armoured truck half off the tracks. The naval gun tried to engage the enemy but was knocked off its mountings. While Haldane and some soldiers sought to keep the Boers pinned down with rifle fire, Churchill led a group of volunteers to decouple the engine from the derailed carriages. 'Nothing was so thrilling as this', wrote Churchill,[49] and they enabled the engine to carry many of the wounded and some soldiers back to Frere, whereupon the Boers captured fifty-six prisoners, including Churchill and Haldane. Churchill received plaudits for displaying 'the greatest coolness' in superintending the clearing of the line.[50]

Apart from the Kraaipan skirmish, 60 km south-west of Mafeking, armoured trains avoided disasters in Cape Colony. The two trains trapped in Kimberley supported the aggressive defence, often drawing enemy fire while mounted units harassed the Boer positions. Another armoured train led the reconnaissance missions, as Lord Methuen's relief column advanced along the railroad towards Kimberley. Unlike the Estcourt train, this train usually had mounted patrols in support, but whenever it encountered enemy fire, both it and any following trains had to retreat rapidly. Sappers accompanied soldiers in the armoured trucks, and, in the advance upon Graspan, they were repairing a small culvert when they came under fire. As Sapper Champion observed, this was 'a most dangerous job' for the engineers, as 'they have to go first and see the line is all right'.[51]

Faced with relief columns in both colonies and massive reinforcements in Cape Colony, the Boers fell back in retreat. Quite remarkably, they removed the Long Tom gun from its firing position outside Kimberley and moved it on a wagon, probably hauled by thirty oxen, to the railhead at Klerksdorp, from whence it was transported by train to Pretoria. Similarly, another Long Tom was withdrawn in haste from its repair position behind the Tugela lines. As Buller broke through the Boer positions, the gun was withdrawn at night around the defences of Ladysmith until it could be loaded on the last train that left Modder Spruit station on 27 February 1900. On the following day, the Wonderboom Tom was loaded onto the last train that left Elandslaagte station and returned to northern Natal. Although the Boers made a brief stand at Helpmekaar (13 May 1900), they were soon in full retreat and loaded both Long Toms onto a train at Glencoe for the final withdrawal to Pretoria.[52] So the Boers did not simply destroy railway property in their retreats,[53] but made use of the railways wherever possible, and then destroyed lines, tunnels, culverts and bridges to hinder the British as much as they could.

Invasion of the Orange Free State and the Transvaal

Following the capture of General Piet A. Cronjé and his army at Paardeberg (27 February 1900), and the relief of Ladysmith and Kimberley, Kitchener informed the queen about the forthcoming march to Bloemfontein: 'Our difficulty regarding transport for so large a force of mounted men is very great, but though we may have to go with short rations and forage no one will grumble, and once at Bloemfontein by opening the Midland railway lines we shall soon be quite comfortable again.'[54] Railway planning, however, moved somewhat erratically. Initially it took only three days to open the line to Kimberley, as the Western Field Railway Section worked by day, the Midland Field Railway Section at night, and they met a construction party working south from Kimberley at Spytfontein on 19 February 1900. While the Midland Field Section then withdrew to Naauwpoort to be held in readiness for the advance through the Free State, Girouard, several staff officers, and a civilian traffic manager rode to join Roberts, intending to organize railway repairs from Bloemfontein southwards. Kitchener countermanded these orders, and sent everyone back to Naauwpoort, to work northwards from there.[55] Only on 26 February was Major-General Ralph A. P. Clements able to begin his pursuit of the retreating enemy northwards through Colesberg to Norval's Pont, while Major-General Gatacre began his advance from Molteno through Stormberg towards Bethulie.

At the Orange River, an immense waterway that formed the boundary with the Orange Free State, columns found the two railway bridges destroyed at both ends. Surveying the wreckage at Norval's Pont, where the river's breadth was some 457 m, Gunner J. E. Hogg (87th Howitzer Battery) reckoned the Boers must have been 'very clever' in their destruction, while Lance-Corporal R. Shaw (7th Volunteer Battalion, Royal Scots) commended the 'good work' of those Boers who left the five spans of Bethulie bridge 'broken and twisted in the water'.[56] Repairing Norval's Pont became the priority, and it required the Midland Field Railway Section and the Railway Pioneer Regiment to supplement the under-staffed gangs of the CGR. They constructed both an aerial tram and a low-level bridge, restoring traffic to Bloemfontein by 27 May (where they had captured twenty-eight engines, 325 trucks and an undamaged line running south to Norval's Pont and Bethulie). Meanwhile Major Graham Thomson, RE, supervised the laying of a railway diversion over the road bridge at Bethulie, which was still undamaged despite the discovery of 132 cases of dynamite packed underneath the bridge. Although this bridge could not take engines, it bore the weight of loaded trucks, which were hand-shunted across and taken away by engines on the other side. Kitchener was hugely impressed:

> The Engineers have worked very well at the railway bridges at Norvals [*sic*] Pont and Bethulie, which were completely destroyed, and I hope before long we shall be able to get up a good supply of clothing, boots, and comforts, for the troops which are much wanted.[57]

During the month of April 1900, 8,600 loaded trucks brought supplies for 65,000 men and 33,000 animals to Bloemfontein. Although the fourteen days of bridge construction had placed Roberts, as Leo Amery observed, 'in direct rail communication with all the Cape ports',[58] many difficulties remained. Despite the Cape lending engines to supplement those captured in the Free State, engine power was in short supply. Bloemfontein lacked sufficient station accommodation, a problem whose effects were only mitigated by the great congestion of trucks south of the Orange River, where supply trains blocked access to numerous stations and tied up vast amounts of rolling stock. The strains on the system north of the Orange River grew inexorably as the advancing army captured hundreds of miles of railroad but without securing any additional rolling stock. Dutch railway officials proved unwilling to co-operate with the invaders, and so Girouard had to create an entirely new railway system, the 'Imperial Military Railways' (IMR). Emulating the arrangements imposed on the Cape railways, this military system would ultimately cover 'the whole of the 1,310 miles [2,108 km] of Boer line'.[59] Girouard found replacement staff from men loaned by the CGR, English staff from the Free State railways, officers and men from the four railway companies of Royal Engineers, and some 800 to 1,000 army reservists and men from the colonial, militia and volunteer units, who had worked on the railways in civilian life and were induced to join by the offer of working pay at the rates of Royal Engineers. Finally, Girouard had to limit the damage from the dynamiting skills of Major John MacBride's Irish Transvaal Brigade, which had laid waste the railway fabric of the Free State, destroying stations, telegraphs, water supplies, permanent way and virtually every bridge between Bloemfontein and the Vaal River.[60]

The construction team under Lieutenant H. A. Micklem, RE, assisted by 'thousands of black railway navvies',[61] could not keep pace with the army, despite a ten-day halt after the capture of Kroonstad to allow time for railway repairs. Advance units pressed ahead, partially in the hope of capturing some rolling stock.[62] They crossed the Vaal River at Viljoen's Drift on 27 May, and seized Johannesburg four days later. Following trains took another ten to twelve days before they reached the Vaal, where the massive bridge with seven spans, each of some

40 m, had been utterly destroyed. At a time when only eight trains per day could reach the front from the coastal ports, one of which carried coal for the railways themselves, and another material necessary for repairs, the condition of the Vereeniging mines was crucial. Fortunately for the British, the Boers, unlike their Irish commandos, were reluctant to destroy valuable property, and so the Irishmen were banned from dynamiting the mines (and railway and telegraph installations).[63] While Lieutenant H. L. Pritchard's repair team used timber from the mines to build a temporary bridge across the Vaal, the railways used its coal to sustain the railway operations.[64] Kitchener, nonetheless, was painfully aware of 'the great difficulty experienced in getting up sufficient supplies from the base, as well as the very large number of horses and mules required to replace sick and supply the wants of the increased number of troops that we now have'.[65] As he explained to the queen, 'With only one narrow gauge line of railway and a limited amount of rolling stock it is not easy to arrange to get up the vast amount that is required.'[66]

Despite the destruction of parts of the railway and more bridges, movement through the Transvaal was relatively rapid, as President Paul Kruger and the Transvaal government fled Pretoria two days before the fall of Johannesburg. More rolling stock fell into British hands: seven engines and 600 trucks at Johannesburg, eight engines and 200 trucks at the railway junction of Elandsfontein and another sixteen locomotives and 400 trucks when Pretoria, the capital of the Transvaal, fell on 5 June. Of all these captures, Elandsfontein, the busiest traffic centre in war or peace, was critical from a railway perspective: as Girouard claimed, 'the seizure of such a station in full working order was of commanding importance', especially as the new bridges across the Vaal and Irene, 16 km south of Pretoria, were ready on 11 and 19 June respectively.[67] Even so, Kitchener worried about 'a very long line of communications and the activity of the Free State Boers' and yearned for support from Buller's army in Natal.[68] Only on 4 July 1900 did units of Roberts's and Buller's armies meet for the first time, and, on 26 July, the railway reopened from Durban to Pretoria. As Lord Roberts observed, 'This will be of great assistance in the matter of supplies as the distance to Durban is only 511 miles [822 km] against 1040 miles [1674 km] to Capetown.'[69] It would also prove a timely acquisition, as Roberts's primary supply line was coming under sustained attack.

Guerrilla war

Far from proving decisive, the occupation of Pretoria had enabled the Transvaal forces to retreat along the Delagoa Bay Railway into the eastern Transvaal, while Free Staters launched all manner of attacks against the IMR system. On 7 June 1900, only two days after the fall of Pretoria, forces under General Christiaan De Wet attacked Roodewal station about 50 km north of Kroonstad. They cut the telegraph and destroyed the station, a mail train, nearby bridges and vast quantities of mail; they also seized winter clothing, 600 cases of ammunition, artillery shells and other supplies – the largest haul of booty in the entire war.[70] An outraged Lord Roberts retaliated by issuing his Proclamation No. 5 on 16 June (and three days later the harsher Proclamation No. 6), declaring that farmsteads in the vicinity of attacks on railway bridges and culverts would be burnt down – the beginning of the British scorched-earth policy.

As with reprisals applied in Afghanistan and India, the aims of the policy were 'strictly military': to punish the culprits and deter others by harassing the civilian population. This included collective punishments, the burning of farms, the removal or killing of livestock and the corralling of homeless women and children in 'refugee camps', later known as concentration camps, often located along the railway lines for purposes of resupply. In September 1900, the radius for the removal or destruction of food supplies was expanded to 25.7 km from an attack site, hugely extending the area of potential devastation. The policy embittered the Boer commandos, relieved them of their family responsibilities, added to their numbers (as many boys under sixteen years and men over sixty years preferred to take up arms rather than go into the camps), and diverted soldiers from combat duties to burning farms, moving civilians and guarding the camps. Attacks on the vulnerable, rail-based supply lines intensified, and, from 7 June 1900 to 28 June 1901, the Boers interrupted the lines on 255 occasions, with the peak of intensity occurring in November 1900.[71]

Although Girouard described these actions as inflicting 'frequent though generally inconsiderable' damage upon the Free State line, he had to suspend the running of night trains between Bloemfontein and the Vaal in July 1900, so reducing the daily supply trains from eight to six. He had to issue similar bans upon night trains between Bloemfontein and the Orange River (and between Pretoria and Waterval) in October, and then throughout the entire IMR system in January 1901.[72] So the Boer attacks were hardly inconsequential, yet traffic resumed across much of the IMR system: the first train reached Mafeking on 9 June, and in the same month the branch line from Johannesburg to Klerksdorp was opened. Girouard also supported Roberts with several large-scale troop movements to counter the guerrilla operations in the Free State, renamed as the Orange River Colony (ORC), and beyond. In a five-day period during July 1900, thirty-one trains conveyed Lord Methuen's mounted force by rail from Kroonstad to Krugersdorp (while sustaining three supply trains per day from the south to Pretoria). In August 1900, when De Wet broke through at Magaliesberg and tried to work round north of Pretoria, parts of Ian Hamilton's and Baden-Powell's mounted forces were railed up to Warmbaths station, where they headed De Wet off. During December 1900 another foray by De Wet in the south of the Free State required the co-ordinated movement of fifty-two troop trains from Bloemfontein across lines in the south of the ORC and the north of Cape Colony. When De Wet actually crossed the Orange River in February 1901, eighty-nine troop trains left Bloemfontein in response, carrying 9,000 men, 14,500 horses, 48 guns and transport. Finally, railways helped to prevent another invasion of Natal in October 1901, moving thousands of men and horses, with guns and transport, to threaten an incursion over a period of thirty-six days.[73]

The other major railway offensive was the eastward advance along the Delagoa Bay railway line, launched on 23 July 1900 and completed on 24 September 1900. Units had already prepared the way as far as Bronkhorstspruit, where the construction train arrived on 25 July. Four supply trains, and then another four, followed, enabling Kitchener to form a depot for the advancing troops. After construction teams completed deviation works (6 August) to offset the blown-up bridges at Bronkhorstspruit and the Wilge River, the advance resumed and profited from the failure of the prepared charges to destroy the Olifants River Bridge. Once British forces had fought and won the last set-piece battle of the war at Bergendal (21–7 August 1900) and had concentrated in the town of Belfast, the remaining advance was by foot or horse; 'not a single troop unit was carried by rail'.[74]

This phase of the advance yielded more rolling stock at Barberton, Kaapmuiden, Avoca and Krokodilpoort. Although some of the trucks were burnt, and the engines often lacked vital parts, the usable material proved helpful in the repair of several bridges. Sapper C. Bowden wrote a vivid account of the work undertaken by his unit, 12th Company, RE, since leaving Molteno. 'We had marched 800 miles [1,288 km] during the last six months, marching with rifle, ammunition, blanket, rations, and sometimes with pick and shovel, etc., in a blazing sun in the day, and lying down on the veldt in the bitter cold at night, often very frosty.' They had fought in skirmishes in the Free State, and at Bergendal; they had built bridges, redoubts and roads; they had trudged ahead of the army, 'making roads through drifts and many other jobs'; and, on the final day of this march, 'we covered 19 miles [31 km], the whole way of which had to be cut through bush and jungle'. Now at Komati Poort, popularly known as 'the white man's grave', the 12th Company were 'alive yet, building huts and making roads with the heat 110 degrees in the shade all through the fever months'.[75]

From 27 September onwards Girouard was able to operate communications by rail from Cape Town to the Portuguese border, a distance of 2,144 km. He had access to another 103 locomotives and 2,645 trucks and carriages found at Komati Poort (see Figure 12) and in the adjacent Portuguese territory, albeit in varying states of repair (some 500 trucks were burnt and 115 of these totally destroyed). The recovery of this material was all the more important because the operation of the IMR network at this time required the regular use of 3,000 trucks and fifteen engines from Cape Colony. 'So straightened was the situation', recalled Girouard, 'that the loss of two engines one on the Krugersdorp line and one near Kroonstad was acutely felt . . .', and new rolling stock, ordered from Britain, had to be brought into service on the C.G.R. and I.M.R. lines, but 'most of it not before October 1900'.[76]

However welcome the rail-based support, the British army committed vast resources to defending and deterring attacks on railway property.

Figure 12 'The Return of the 1st Scots Guards from Komati Poort: Emergency Motive Power', *Illustrated London News*, vol. 117, no. 3216 (8 December 1900), p. 850

It was demoralizing work: initially the railway defences consisted of merely open and inconspicuous trenches at stations, bridges and culverts, with men undertaking mounted and foot patrols along the line. 'We are all tired of this life, and wish it was over', wrote Private D. K. MacKenzie (Seaforth Highlanders) while guarding a railway bridge near Heilbron. He had almost starved, living 'on Indian corn and seed meal ground down and made into porridge . . . black and dirty looking'. He deeply resented the burning of the mail at Roodewal, 'and me expecting a letter from my girl', and found it 'awfully cold' in mid-winter, with the water frozen at night 'in our water bottles, and we have no tents, only two thin blankets'.[77] Writing from Olifantsfontein, Corporal Libby described how

> Our principal duties by night are outposts and patrols. Every night we have to patrol near the railway until we get in to touch with other companies. One of the patrols means about fourteen miles [22.5 km] and the other eight [13 km], and walking along a railway that distance we find quite enough.[78]

As the British and their black guides became more proficient in detecting suspicious footprints or displaced stones wherever a mine was laid or a length of rail unbolted, the Boers, as Roland W. Schikkerling admitted, had to adapt their techniques. At night two men, he explained, would carry a shortened Martini Henry rifle loaded with a blank cartridge, dynamite and a bag, and 'holding hands' for about a hundred yards [91 m] would 'walk along the metals to avoid leaving footprints'. The aim was to place the shortened rifle, with trigger guard removed, upside down under one of the rails, and cover it with gravel, so that when the rail bent under the weight of a passing locomotive, the gun fired into some dynamite cartridges placed in its mouth, tearing 'the rail asunder and the train falls over or is stranded'. After covering the dynamite, and putting all displaced stones in the bag, 'the men again walk back for some distance along the metals'.[79]

The Boers were at their most disruptive, wrecking trains, destroying bridges, cutting telegraph lines and damaging the permanent way, when Kitchener assumed overall command in South Africa on 29 November 1900. The following months were among the worst of the war, especially along the Pretoria–Delagoa line, where Boers under General Ben Viljoen, Jack Hindon and Carl Trichardt, proved highly destructive. Kitchener, who was twice nearly captured while travelling by train,[80] faced an acute dilemma. As he explained to William St John Brodrick, the new secretary of state for war, he had to protect the 'very long lines of railway & road' and supply garrisons all over the country, which left him with 'no striking force of any importance' in case of emergency.[81] His first response was to rationalize the number of outlying garrisons. By 18 January 1901, Kitchener reported that he had brought in garrisons from Phillipolis, Jagersfontein, Smithfield and Rousville, and so 'the garrisons of the two latter are now acting as a mobile column in the Zastron district'.[82] He also intensified the policies of farm-burning, livestock destruction and concentration camps begun by his predecessor.

Ultimately the British built over forty camps for whites, holding some 116,000 inmates by the end of the war, and another sixty camps for 115,000 black inmates. Although conditions in the camps varied considerably, depending upon the policies of the superintendents as well as the facilities and supplies available, their management and mounting death toll (eventually 28,000 Boers died, 22,000 of whom were under sixteen years of age), encountered criticisms from Emily Hobhouse among others. Many soldiers, nonetheless, supported the policy: 'clearing off the people', wrote Sergeant William Hamilton (Highland Light Infantry), 'has destroyed their principal means of intelligence'.[83] Other soldiers, surviving on half-rations on the veld, defended the camps, because the inmates were 'better treated than our own men' or 'probably better than many of our own men's wives at home'.[84] Kitchener was certain the camps were 'doing good work'. As he assured Brodrick,

> The women left on farms give complete intelligence to the boers of all our movements and feed the commandos in their neighbourhood. Where they are brought in to the railway they settle down and are quite happy they even may give some intelligence but it is very little . . .[85]

He sought, too, to upgrade the railway defences by replacing the open trenches with the closed works that had a wide field of fire and additional protection, involving barbed wire and other obstacles. While some of these buildings were substantial stone forts to guard railway bridges, others were at first small octagonal structures, made of two skins of corrugated iron nailed onto wooden frames. They needed regular foundations and loopholes drilled in the centre, and proved costly and time-consuming to erect. In February and then March 1901, Major Spring E. Rice, RE, simplified this design, first by reducing the width between the sheets of corrugated iron and by removing the need for a foundation, and then by developing an even cheaper and more efficient option. This was a blockhouse based upon two cylinders of corrugated iron without any need for woodwork because the space between the two sheets could be packed with shingle. The new blockhouse held a garrison of seven men. Roofed and loopholed as before, the whole structure was relatively cheap, costing £16 apiece when made by Royal Engineers. It was easily erected, requiring relatively little material, transport or skilled labour, and proved more durable than its predecessors.

By May 1901, most of the line from Pretoria to Komati Poort was well protected, with blockhouses stationed at regular intervals 'of about a mile and a half [2.4 km] down the whole extent of the line', and this interval was steadily lessened, becoming 'as small as 400 yards [366 m] on the Delagoa line' and '200 yards [183 m] on portions of the Cape railways. A continuous fencing of barbed wire eventually ran along the line; elaborate entanglements surrounded each blockhouse, and the telephone linked up the whole system.'[86] Trenches bordering the barbed wire and running to within 91 m of each blockhouse were added, and tins suspended on sticks gave a tinkling whenever anyone approached or tried to pass through the entanglement at night. By July 1901, the army had erected some 8,000 blockhouses over a total length of 5,955 km; Kitchener reassured Brodrick that 'I have been for some time fortifying the railway lines with blockhouses so as to reduce the numbers employed in defending the lines, which duty takes up by far the greater number of troops in S.A.'[87]

Coupled with the blockhouse system was the revival of the armoured train in a much more advanced design than the previous versions. From front to rear the No. 3 train, introduced in January 1901, included a truck carrying rails, sleepers and repair materials; an armoured truck equipped with a Maxim machine gun; a large 'bogie' truck containing a naval 12-pounder with a range of 7,315 m; a truck containing a dynamo and oil engine to work the searchlight that could be set up at either end of the train as required; the engine armoured with sheet metal and capable of running at 80.5 km an hour under favourable conditions; a truck fitted out with supplies such as water, milking goats and a poultry run; a truck divided between the enginemen and the canteen; an officer's carriage, the far end of which was divided between the sergeants and telegraphists; and finally another armoured truck similar to the one in front. Subsequently the whole train was lit with the aid of electric lamps borrowed from hospital trains.[88]

The refurbished armoured trains served as escorts to the construction trains in December 1900, and at least nineteen were brought into service because their potential utility soon became apparent. Intended to keep the railway lines free for an unobstructed flow of traffic, they were used for patrolling, reconnaissance and reinforcement of points in the line, and as offensive weapons in conjunction with columns in the field. Effectively they became mobile forts (see Figure 13), with Girouard claiming that 'the enemy disliked them intensely, and the presence of an armoured train had a great moral effect'.[89] These armoured trains certainly provided reassurance for the soldiers whom they escorted,[90] as well as relief for those who found themselves ambushed along the line before an 'armoured train arrived', whereupon 'a few rifle shots and a rat-tat of the pom-pom disposed of the loitering Boers'.[91] They were also disruptive, as General Ben Viljoen and his commandos found when they tried to storm two blockhouses and cross the Delagoa Bay railway line at Belfast on 27 June 1901.[92]

Manning the armoured trains was a lively and challenging experience, as the soldiers of the Highland Light Infantry found on Armoured Train No. 3. They had varied work, clearing obstacles off railway lines,

Figure 13 'The Hemming-In of the Boers: An Armoured Train Foiling an Attempt of Burghers to Cross the Railway' (drawn by R. Caton Woodville), *Illustrated London News*, vol. 120, no. 3275 (25 January 1902), p. 117

dispersing Boers as they were looting blown-up trains, harrying De Wet's column in Cape Colony, co-ordinating operations with mounted infantry against Boers in the vicinity of railway lines and later participating in the major 'drives' that dominated the final phases of the war. In these 'drives' the trains travelled up and down sections of the railway line with searchlight displays at night, and then periodically engaged in long-range pounding of Boer parties, sometimes 'at 5,600 yards [5.1 km]', with their 12-pounder guns during the day.[93]

None of these effects would have been possible had the armoured trains been left under officers commanding particular sections of the line. This was the original arrangement, but because it increased traffic congestion, the trains were placed under the overall direction of Captain H. C. Nathan, RE. A former deputy assistant director of railways at Kimberley until May 1900, he held that position at Kroonstad for several months, after which he commanded an armoured train (and at Baartman Siding captured much of De Wet's ammunition and supplies). He was thereby both acquainted with the regulations of the railway and the handling of armoured trains. Given a broad remit, Nathan was expected to co-ordinate the movements of the armoured trains by liaising with army headquarters and railway managers. Allowed to experiment with different designs of trains, he was responsible for their distribution and for keeping them equipped and armed. Nathan also instructed officers in command of the armoured trains about the best tactics, the methods of patrolling and their working with local railway officials. Although the armoured trains, once deployed, remained under the orders of generals or officers commanding particular sections of the line, they could be moved by Nathan whenever he thought that they might be needed elsewhere.[94]

Operationally, the challenge remained formidable: 'The enemy', wrote Kitchener, 'are in scattered parties of from 30 to 200 over an immense area, and sometimes in most difficult country, with which they are naturally thoroughly acquainted . . . They have excellent scouts and information and generally move away rapidly before our columns can get within 20 miles [32 km] of them.'[95] In these circumstances railways could sustain operations, and could influence local outcomes where Boers were prevented from crossing blockhouse lines,[96] but the British and imperial forces still had to operate more creatively on the veld. Boer joiners refined their scouting techniques; by the end of the war, 5,464 of them served the British, including Transvaalers in the National Scouts and Free Staters in the Orange River Colony Volunteers, under the command of General Piet De Wet (brother of Christiaan).[97] Blacks served, too, as labourers, cattle guards, guides, messengers and blockhouse watchmen, and at times proffered armed assistance. By mid-March 1902, Kitchener acknowledged that 4,618 blacks were working as watchmen on the railways and blockhouse lines in Natal, Orange River Colony and the Transvaal.[98] More significantly, he recognized that British and imperial forces were learning to conduct their 'drives' quite differently. As he wrote on 16 August 1901, some of the 'mobile columns scouring the country in every direction . . . are specially fitted to travel light and fast'.[99]

While these 'drives' rounded up variable numbers of burghers, arms, wagons and livestock, and occasionally captured a leading general (Pieter H. Kritzinger on 16 December 1901 and Ben Viljoen on 25 January 1902), Boers were more impressed by the enemy's new methods of warfare, particularly the improvements in British tracking, raiding and night operations.[100] The Boers still crossed the railway lines, triumphing in several engagements at Bakenlaagte (30 October 1901), Tweefontein (25 December 1901) and Tweebosch (7 March 1902), and General Jan C. Smuts led a commando that reached within 241 km of Cape Town by the end of February 1902. Their attacks on the IMR network, though, petered out entirely in October–December 1901, and only isolated incidents recurred in the early months of 1902. Girouard attributed this success to the extension of the blockhouse lines, with barbed wire and alarms more extensively deployed than ever before. Ultimately the blockhouse system was, in his words, 'a practical solution which answered admirably'.[101] Quite apart from armoured trains operating in support of various 'drives', rail-based support for the 50,000 men and the thousands of black auxiliaries on blockhouse duty involved an intricate system of water trains in the dry season, ration trains every two days and supply trains every twelve days to sustain the garrisons with food, water, rum, ammunition, mail and other supplies. If work on the blockhouse line was desperately dull and tedious, it was remarkable that in May 1902 a newly arrived soldier, from the 3rd Battalion, Leicestershire Regiment, could describe his blockhouse existence as 'a toff's life'. Spared trudging across the veld on half-rations, Private Ernest Griffin reckoned that he had 'no work to do but to sleep and eat', and that the food was 'very good' and 'plenty of it'.[102]

As attrition took its toll, and peace talks began at Klerksdorp, the last battle of the war took place at Rooiwal (11 April 1902), where a massed, mounted charge by the Boers was broken up with heavy losses. A peace conference met in Vereeniging, and safe conduct was guaranteed for all the Boer leaders due to attend. Deneys Reitz, who accompanied General Jan C. Smuts from Cape Colony, remembered his excitement at 'travelling by rail for the first time for nearly two years. . . . After years of rough fare and hard living, we had luxurious cabins, with soft beds to lie on; a steward with coffee in the morning, a bath ready prepared and food such as I had almost forgotten the existence of.' After five days he reached Cape Town and eventually received orders to travel north. From Magersfontein they continued their journey at night only, with 'an armoured train puffing ahead all the way, its searchlight sweeping the veld . . . I have been told that we were purposely delayed lest, coming from the Cape where the outlook was brighter, we might persuade the Transvaalers that things were not as bad as they seemed.'[103] A deliberate plot is probably doubtful, but the use of a railway train to convey images of civilized living in a post-war environment can only have boosted Kitchener's chances of securing a negotiated settlement.

Railways remained critically important throughout the South African War. Both sides recognized their importance and exploited them initially, as all the principal towns, junctions and mines were located close to the extensive rail network. The single-track system was by no means ideal but it permitted a multitude of tactical purposes including reconnaissance, the transport of men, horses, ordnance and supplies, fire support and the removal of wounded and prisoners from the battlefield. The rail network, too, served as primary means of social interaction. It enabled colonists in Natal and Cape Colony to demonstrate their support for the British soldier, facilitated meetings with an enemy, who was often unseen on the battlefield or too elusive on the veld, built up new communities at the enlarged railway stations and engaged the services of thousands of black workers willing to assist in repairing rail lines, stations, culverts and bridges, and then later to act as guides and watchmen on the blockhouse lines.

Operationally, British reinforcements had to use railways if they were to move rapidly into the interior to relieve beleaguered towns, and later to try and prevent invasions of Natal and Cape Colony. Although they were not always successful, the British and imperial forces could never have responded to Boer movements had they not developed a system of traffic management and railway control under Girouard's direction. Roberts may have seized the strategic initiative when he left the railway line and advanced on Bloemfontein, but he could never have made this manoeuvre, nor exploited his subsequent advance, without a massive movement of men, guns, horses and supplies by rail in the first instance, and then without further support when rail connections were restored across the Orange River. By lengthening his lines of communication Roberts only placed a higher premium upon the protection of his rail-based support, the reopening of the Pretoria to Durban line, and the reconstituting and operation of the captured IMR network.

Kitchener as an engineer, who had been applauded for his rail-based operations in the Sudan, understood the significance and vulnerability of his line of communications in South Africa. He knew that sustaining a functioning railway in spite of the recurrent attacks added to the cost of an increasingly expensive war but reckoned correctly that he could neither maintain the counter-insurgency nor meet the needs of a dependent civilian population without the railway.[104] Protecting the railway, and then using it both defensively and offensively, proved critical components of an attrition strategy that ultimately prevailed. Despite the depleted rolling stock, and the Boer attacks on the railway system, Girouard and his staff managed to keep the railways functioning and provided invaluable support for the 250,000 men[105] deployed across an inhospitable terrain that exceeded the size of France and Spain. The British army had never before undertaken an imperial mission of such magnitude, involving an unprecedented volume of logistical support protected by a massive blockhouse and barbed wire system.

The price paid for this achievement included counter-guerrilla tactics that proved highly controversial at the time and subsequently. Neither the scorched-earth operations nor the recourse to concentration camps proved decisive in themselves, although the effects of both were among the factors that persuaded fifty-four of the sixty Boer leaders to acquiesce in the final peace settlement.[106] Grinding an adversary down in a war of attrition was not simply a function of keeping the enemy isolated from external assistance through the Royal Navy, and of bringing to bear much larger numbers of British and imperial forces, but it was also a matter of adopting new methods of warfare, of refining military instruments such as the armoured train, and of finding ways to manage and co-ordinate operations across a vast and inhospitable land mass. Utilizing the railways underpinned many of these endeavours.

Notes

1 I. F. W. Beckett, 'Military High Command in South Africa, 1854–1914' in P. A. Boyden, A. J. Guy and M. Harding (eds.), *Ashes and Blood: The British Army in South Africa, 1795–1914* (London: National Army Museum, 2001), pp. 60–71, at p. 68; Col. Sir C. M. Watson, *History of the Corps of Royal Engineers* (Chatham: Institution of Royal Engineers, 1934), vol. 3, pp. 104–5; B. Roberts, *Kimberley: Turbulent City* (Cape Town: David Philip, 1976), p. 236.

2 'Single Lines of Railway in Warfare', *Newcastle Daily Chronicle*, 3 May 1900, p. 5.

3 W. Baring Pemberton, *Battles of the Boer War* (London: Pan Books, 1964), p. 36.

4 D. Porch, *Wars of Empire* (London: Cassell, 2000), p. 120.

5 I. van der Waag, 'South Africa and the Boer Military System' in P. Dennis and J. Grey (eds.), *The Boer War: Army, Nation and Empire. The 1999 Chief of Army/*

Australian War Memorial Military History Conference (Canberra: Army History Unit, 2000), pp. 45–69, at p. 53. For another catalogue of Boer destruction, see Maj. B. A. Zurnamer, 'The state of the Railways in South Africa during the Anglo-Boer War 1899–1902', *Scientia Militaria: South African Journal of Military Studies*, vol. 16, no. 4 (1986), pp. 26–33, http://scientiamilitaria.journals.ac.za (accessed 18 April 2014).

6 TNA, WO 108/378, Lt.-Col. Sir E. P. C. Girouard, *History of the Railways during the War in South Africa, 1899–1902* (London: HMSO., 1903); L. S. Amery (ed.), *The Times History of the War in South Africa 1899–1902*, 7 vols. (London: Sampson, Low, Marston & Co., 1900–9), vol. 6, ch. 3; F. Pretorius, *Historical Dictionary of the Anglo-Boer War* (Lanham, Maryland: Scarecrow Press, 2009), pp. 362–5.

7 Wolmar, *Engines of War*, p. 102.

8 Pretorius, *Historical Dictionary*, p. 12.

9 'In Beleaguered Kimberley', *Birmingham Daily Post*, 20 March 1900, p. 7; see also E. M. Spiers (ed.), *Letters from Kimberley: Eyewitness Accounts from the South African War* (Barnsley: Pen & Sword, 2013), ch. 1.

10 Pretorius, *Historical Dictionary*, pp. 362–3; L. Changuion, *Silence of the Guns: The History of the Long Toms of the Anglo-Boer War* (Pretoria: Protea Book House, 2001), pp. 14–16 and 26.

11 'The Battle of Elandslaagte', *Tamworth Herald*, 9 December 1899, p. 5; see also E. M. Spiers (ed.), *Letters from Ladysmith: Eyewitness Accounts of the South African War* (Barnsley: Pen & Sword, 2010), ch.1.

12 'A Soldier's Letter', *Liverpool Mercury*, 14 December 1899, p. 8.

13 Nurse Kate Driver, *Experience of a Siege: A Nurse Looks Back on Ladysmith*, Diary the Siege of Ladysmith, 6, revised edition (Ladysmith: Ladysmith Historical Society, 1994), p. 4.

14 'Letter from Pretoria', *Cheltenham Chronicle*, 16 December 1899, p. 5.

15 'The Experiences of Local Outlanders [*sic*]', *Newcastle Daily Chronicle*, 15 November 1899, p. 8.

16 'In Camp at De Aar', *Birmingham Daily Gazette*, 28 October 1899, p. 5; see also 'A Letter from Ladysmith', *Aberdeen Free Press*, 4 December 1899, p. 6, and D. Cammack, *The Rand at War, 1899–1902: The Witwatersrand and the Anglo-Boer War* (London: James Currey, 1990).

17 'A Bristolian's Experiences in Bloemfontein', *Bristol Observer*, 12 May 1900, p. 2.

18 'Letters from the Front. A Cornish Reservist's Experiences', *Western Morning News*, 16 January 1900, p. 8; see also 'A Boer Boy Warrior', *Manchester Evening News*, 17 January 1900, p. 5.

19 'The Gordons at Ladysmith', *Aberdeen Free Press*, 15 November 1899, p. 6, and 'More Letters Home', *[Gloucestershire] Echo*, 25 November 1899, p. 4.

20 'News from Somersets at the Front', *Somerset County Gazette*, 30 December 1899, p. 9; see also Private Tucker, diary, 25 November 1899, in P. Todd and D. Fordham, *Private Tucker's Boer War Diary* (London: Elm Tree Books, 1980), p. 22.

21 G. Powell, *Buller: A Scapegoat?* (London: Leo Cooper, 1994).

22 E. M. Spiers, 'Intelligence and Command in British Small Colonial Wars of the 1990s', *Intelligence and National Security*, vol. 22, no. 5 (2007), pp. 661–81; Pretorius, *Historical Dictionary*, pp. 90–2, 259–62, 285–8 and 434–6; Spiers, *Letters from Ladysmith*, pp. 55–73, and *Letters from Kimberley*, pp. 42–75, 81–103.

23 TNA, WO 108/378, Girouard, *History of the Railways*, p. 25.

24 Col. J. Y. F. Blake, *A West Pointer with the Boers* (Boston: Angel Guardian Press, 1903), p. 157; see also A. Wessels, 'Afrikaners at War' in J. Gooch (ed.), *The Boer War: Direction, Experience and Image* (London: Frank Cass, 2000), pp. 73–106, at pp. 82–5, and Pretorius, *Historical Dictionary*, pp. 125–6.

25 TNA, WO 108/378, Girouard, *History of the Railways*, p. 15; see also TNA, CAB 17/11, Committee of Imperial Defence, 'Railway Organisation in Theatres of War, 1903', G. F. Ellison, 'Précis of Views of Lt.-Col. Sir Percy Girouard, KCMG, in his Interview with Lord Esher, 26 November 1903'.

26 TNA, WO 108/378, Girouard, *History of the Railways*, pp. 15 and 30; see also G. F. Williams, *The Diamond Mines of South Africa: Some Account of their Rise and Development* (New York: Macmillan, 1902), p. 626.
27 TNA, WO 108/378, Girouard, *History of the Railways*, pp. 25 and 20; Amery (ed.), *Times History*, vol. 6, p. 313.
28 TNA, WO 108/378, Girouard, *History of the Railways*, pp. 16 and 32.
29 'Letters from the Theatre of War: Why the Boers did not Attack De Aar', *Western Morning News*, 29 November 1899, p. 8. On the drink ban initially, see 'Going to the Front', *Birmingham Daily Gazette*, 23 October 1899, p. 4.
30 'The Tension of Waiting: Affairs at De Aar', *Western Morning News*, 18 November 1899, p. 8.
31 'The War Letter', *Morning Leader*, 14 November 1899, p. 6, and 'In Camp at De Aar'.
32 'Letters from the Front', *Gloucester Journal*, 20 January 1900, p. 7; on the sand storms, see C. Miller, *Painting the Map Red: Canada and the South African War, 1899–1902* (Montreal and Kingston: Canadian War Museum and McGill-Queen's University Press, 1993), p. 77.
33 'The Camp at De Aar', *Liverpool Mercury*, 15 February 1900, p. 7.
34 W. Hart-McHarg, *From Quebec to Pretoria: With the Royal Canadian Regiment* (Toronto: William Briggs, 1902), p. 59.
35 Ibid. pp. 59–60.
36 'The Spirit of the Gloucesters', *Gloucestershire Chronicle*, 23 December 1899, p. 7, and 'Home from the War', *Stroud Journal*, 10 August 1900, p. 3. The Cape railways treated the soldiers consistently better than their Natalian counterpart:, Sir G. Fleetwood Wilson (Qs. 5996 and 6257), in evidence appended to PP, *Report of Her Majesty's Commissioners Appointed to Inquire into Military Preparations and other Matters Connected with the War in South Africa*, Cd 1789 (1904), XL, and TNA, WO 108/378, Girouard, *History of the Railways*, p. 15.
37 'Letters from Swindonians at the Front', *Evening Swindon Advertiser*, 13 June 1900, p. 3.
38 'Another Letter from the Front', *Dorset County Chronicle*, 19 April 1900, p. 12; Lt.-Col. C. K. Wood, 'The Work of the Royal Engineers in Natal', *Professional Papers of the Corps of Royal Engineers*, Occasional Papers, 27 (Chatham: Royal Engineers Institute, 1901), pp. 49–70; and Tucker, diary, 10 June 1900, in Todd and Fordham, *Private Tucker's Diary*, p. 113.
39 The 2nd Cheshire Royal Engineers (Railway) Volunteer Corps, first formed in 1887, included six companies of over 100 men apiece. By 1899, there were 760 men in the corps, 245 of whom were classed as regular reservists, having enlisted in the regular army for a single day and so incurring a liability to return to active service at any time over the next six years. These men, along with another two detachments (40 men), were mobilized immediately for war. 'The Crewe Engineers for the Transvaal', *Crewe Guardian*, 18 October 1899, p. 3.
40 'Crewe Engineer Reservists in South Africa', *Crewe Guardian*, 20 January 1900, p. 4.
41 'Modder River Bridge, a Sight', *Crewe Guardian*, 20 January 1900, p. 4.
42 Although soldiers drank from the Modder River, the water was contaminated and so water-borne diseases, known as the 'Modders', spread rapidly: 'Doctors at the Front', *Daily News*, 7 May 1900, p. 3.
43 'An Exonian's Letter', *Devon Weekly Times*, 30 March 1900, p. 7.
44 'A Loughborough Man with French's Column', *Leicester Chronicle*, 24 February 1900, p. 3.
45 'Royal Engineers at Work', *Newcastle Daily Chronicle*, 2 January 1900, p. 5.
46 'More Letters from the Seat of War', *[Doncaster] Chronicle*, 12 January 1900, p. 6.
47 'Among the Wounded', *Cork Constitution*, 20 January 1900, p. 6.
48 'Letter from Sir William MacCormac on the Wounded after the Battle of Colenso', *Fife News*, 20 January 1900, p. 3; see also TNA, WO 108/378, Girouard, *History of the Railways*, p. 55, and Lt.-Col. John H. Plumridge, *Hospital Ships and Ambulance Trains* (London: Seeley, Service & Co., 1975), ch. 8.

49 W. S. Churchill, *London to Ladysmith via Pretoria* (London: Longmans, Green & Co., 1900), p. 89.
50 'The Capture of the Estcourt Train', *Newcastle Daily Chronicle*, 11 December 1899, p. 5; 'The Armoured Train DisasterL Official Report' *[Gloucestershire] Echo*, 20 November 1899, p. 3; Maj.-Gen. Sir F. B. Maurice and M. H. Grant, *History of the War in South Africa 1899–1902*, 4 vols. (London: Hurst & Blackett, 1906–10), vol. 1, p. 268.
51 'Crewe Engineer Reservists in South Africa', p. 4; and on Kimberley, see Spiers, *Letters from Kimberley*, pp. 20, 28–9, 31–2.
52 Changuion, *Silence of the Guns*, pp. 81, 87–8, 91 and 93.
53 Wolmar, *Engines of War*, p. 103.
54 RA, VIC/MAIN/P/7/38, Queen Victoria's MSS, Lord Kitchener to Queen Victoria, 4 March 1900.
55 TNA, WO 108/378, Girouard, *History of the Railways*, p. 29.
56 'In Camp at Norval's Pont', *Coventry Herald*, 27 April 1900, p. 8; 'With the Royal Scots Service Company', *Haddingtonshire Advertiser*, 20 July 1900, p. 2.
57 RA, VIC/MAIN/P/7/242, Queen Victoria's MSS, Lord Kitchener to Queen Victoria, 30 March 1900; TNA, WO 108/378, Girouard, *History of the Railways*, pp. 29–30; and on the discovery of dynamite at Bethulie, see 'From East London to Komati Poort', *North Devon Herald*, 23 May 1901, p. 3.
58 Amery (ed.), *Times History*, vol. 6, pp. 314–15.
59 Ibid. vol. 6, p. 316.
60 TNA, WO 108/378, Girouard, *History of the Railways*, p. 41; see also D. P. McCracken, *MacBride's Brigade: Irish commandos in the Anglo-Boer War* (Dublin: Four Courts Press, 1999), pp. 114–16, and T. Pakenham, *The Boer War* (London: Weidenfeld & Nicolson, 1979), p. 422.
61 Pakenham, *Boer War*, p. 422.
62 Ibid.; Lord Roberts to Lord Lansdowne, 30 May 1900, in A. Wessels (ed.), *Lord Roberts and the War in South Africa 1899–1902* (Stroud, Gloucestershire: Sutton Publishing for the Army Records Society, 2000), p. 86.
63 Enforcing the ban was more difficult but the Irish units lost men through desertion and capture, while most of the brigade retreated east with the Transvaalers and fought in the last set-piece battle of Bergendal (21–7 August 1900). McCracken, *MacBride's Brigade*, pp. 116, 120–1, 126–8, 130–2.
64 TNA, WO 108/378, Girouard, *History of the Railways*, p. 39.
65 RA, VIC/MAIN/P/8/123, Queen Victoria's MSS, Lord Kitchener to Queen Victoria, 26 April 1900.
66 Ibid.
67 Some of the engines had vital parts removed. TNA, WO 108/378, Girouard, *History of the Railways*, pp. 40–1.
68 RA, VIC/MAIN/P/10/36, Queen Victoria's MSS, Lord Kitchener to Queen Victoria, 6 June 1900.
69 Roberts to Lansdowne, 5 July 1900, in Wessels (ed.), *Lord Roberts*, p. 113; Pakenham, *Boer War*, p. 454; Amery (ed.), *Times History*, vol. 6, p. 320.
70 C. R. De Wet, *Three Years War (October 1899–June 1902)* (London: Constable, 1903), ch. 14.
71 Pretorius, *Historical Dictionary*, pp. 365, 396–7; PP, *Army. Proclamations Issued by Field Marshal Lord Roberts in South Africa*, Cd. 426 (1900), LVI, no. 5 of 16 June 1900 and no. 6 of 19 June 1900, pp. 10–11; S. B. Spies, *Methods of Barbarism? Roberts and Kitchener and Civilians in the Boer Republics, January 1900–May 1902* (Cape Town: Human & Rousseau, 1977), pp. 102–3, 110–11, 285, 290, 292–3; Wessels (ed.), *Lord Roberts*, pp. 89, 91–2; Amery (ed.), *Times History*, vol. 6, pp. 329–30.
72 TNA, WO 108/378, Girouard, *History of the Railways*, pp. 42 and 46.
73 Ibid. pp. 44–5; Amery (ed.), *Times History*, vol. 6, p. 328.
74 TNA, WO 108/378, Girouard, *History of the Railways*, pp. 43–4; see also 'The Railway Destruction', *Birmingham Daily Gazette*, 12 October 1900, p. 5.

75 'From East London to Komati Poort', *North Devon Herald*, 23 May 1901, p. 3.
76 TNA, WO 108/378, Girouard, *History of the Railways*, pp. 44, 46, 53–5; Amery (ed.), *Times History*, vol. 6, p. 321; for an account of the negotiations with the Portuguese authorities, see RA, VIC/MAIN/P/14/27, Queen Victoria's MSS, Lord Kitchener to Queen Victoria, 12 October 1900, and for the bureaucratic squabbling about the cost of sending rolling stock from Britain, see A. Page, 'The Supply Services of the British Army in the South African War 1899–1902', unpublished PhD thesis (University of London 1976), pp. 102–3, 339.
77 'Letter from a Seaforth', *Inverness Courier*, 31 July 1900, p. 5; see also 'Corporal Grant, 2nd Gordon Highlanders, Tells Captain Stuart of Inverfiddich his Campaigning Experiences', *Northern Scot & Moray and Nairn Express*, 6 July 1901, p. 3, and Amery (ed.), *Times History*, vol. 5, p. 257.
78 'Letter from Corpl. Libby', *West Briton and Cornwall Advertiser*, 2 August 1900, p. 2; see also 'Guarding the Railway', *Northern Whig*, 9 January 1900, p. 6.
79 R. W. Schikkerling, *Commando Courageous (A Boer's Diary)* (Johannesburg: Hugh Keartland, 1964), p. 157; see also General B. Viljoen, *My Reminiscences of the Anglo-Boer War* (London: Hood, Douglas & Howard, 1903), p. 293.
80 'Hunting De Wet', *Newcastle Daily Chronicle*, 27 February 1901, p. 5; see also De Wet, *Three Years War*, p. 147.
81 TNA, PRO 30/57/22, Kitchener MSS, Lord Kitchener to W. St. J. Brodrick, 20 December 1900.
82 Ibid. Kitchener to Brodrick, 18 January 1901.
83 'Letter from the Front', *Argyllshire Herald*, 6 April 1901, p. 3.
84 'The Boer Refugees', *Bristol Times and Mirror*, 26 July 1901, p. 8, and 'A Notable Letter from the Front', *Ross-shire Journal*, 21 March 1902, p. 5.
85 TNA, PRO 30/57/20, Kitchener MSS, Kitchener to Brodrick, 7 March 1901; see also A. Wessels (ed.), *Lord Kitchener and the War in South Africa 1899–1902* (Stroud, Gloucestershire: Sutton Publishing for the Army Records Society, 2006), p. 100.
86 Amery (ed.), *Times History*, vol. 5, pp. 258–9; D. Judd and K. Surridge, *The Boer War* (London: John Murray, 2002), pp. 214–15.
87 TNA, PRO 30/57/22, Kitchener MSS, Kitchener to Brodrick, 5 July 1901; see also Watson, *History of the Corps of Royal Engineers*, vol. 3, p. 126, and D. W. Aitken, 'The British Defence of the Pretoria–Delagoa Bay Railway', *Military History Journal*, vol. 11 (1999), http://samilitary history.org/vol113da.html (accessed 15 August 2008).
88 'Some Experiences with an Armoured Train in South Africa', *Highland Light Infantry Chronicle*, vol. 3, no. 6 (1902), pp. 743–7; there were different forms of armament on these trains, including light-weight but rapid-firing Colt automatic guns: 'Armoured Train Carrying Colt Guns at Braamfontein', *Illustrated London News*, vol. 117, no. 3216 (8 December 1900), p. 862.
89 TNA, WO 108/378, Girouard, *History of the Railways*, p. 65. As this is the official railway history, I have used Girouard's claim that nineteen armoured trains entered service at this time; other sources claim that twenty were in service.
90 'Gunner Pugsley's Experiences', *Totnes Times*, 20 July 1901, p. 5.
91 'Boer Methods of Warfare', *North Devon Herald*, 21 March 1901, p. 2.
92 Viljoen, *My Reminiscences*, pp. 231–4; Aitken, 'British Defence of the Pretoria–Delagoa Railway', p. 4.
93 'Some Experiences with an Armoured Train', pp. 744–7.
94 TNA, WO 108/378, Girouard, *History of the Railways*, p. 65.
95 RA, VIC/MAIN/W/60/130, King Edward VII MSS, Lord Kitchener to Edward VII, 16 August 1901.
96 'A Cheltonian at the Front', *[Gloucestershire] Echo*, 12 March 1902, p. 3; see also Pretorius, *Historical Dictionary*, p. 202.
97 TNA, CO 417/362/9, 'Return of Ex Burghers Serving under British for Week Ending 1st June 1902'; see also A. Grundlingh, *The Dynamics of Treason: Boer Collaboration in the South African War of 1899–1902* (Pretoria: Protea, 2006).

98 Ministry of Defence (MoD), Whitehall Library, Confidential Telegrams, no. 1022, p. 419, Kitchener to Secretary of State for War, 17 March 1902; W. R. Nasson, 'Africans at War' in Gooch (ed.), *Boer War*, pp. 126–40, and 'Moving Lord Kitchener: Black Military Transport and Supply Work in the South African War, 1899–1902, with Particular Reference to the Cape Colony', *Journal of Southern African Studies*, vol. 11, no. 1 (1984), pp. 25–51.

99 RA, VIC/MAIN/W/60/130, King Edward VII MSS, Kitchener to Edward VII, 16 August 1901.

100 Schikkerling, *Commando Courageous*, p. 207; J. D. Kestell and D. E. van Velden, *The Peace Negotiations* (London: Richard Clay & Sons, 1912), p. 179; E. M. Spiers, 'The Learning Curve in the South African War: Soldiers' Perspectives', *Historia*, vol. 55, no. 1 (2010), pp. 1–17.

101 TNA, WO 108/378, Girouard, *History of the Railways*, p. 67 and appendix D; Judd and Surridge, *Boer War*, pp. 216–17.

102 D. Buttery, 'A "Toff's Life" in the Blockhouse', *Soldiers of the Queen*, no. 103 (2000), pp. 21–3, Private E. Griffin to A. Moore, 25 May 1902, at p. 22; on monotonous life in blockhouses, see 'A Bridgwater Soldier's Letter', *Devon and Somerset Weekly News*, 6 February 1902, p. 2 and 'Life in a Blockhouse', *North Star and Farmer's Chronicle*, 22 May 1902, p. 5.

103 D. Reitz, *Commando: A Boer Journal of the Boer War* (London: Faber & Faber, 1929), pp. 317–19.

104 TNA, PRO 30/57/22, Kitchener MSS, Kitchener to Brodrick, 25 March 1901 and 5 July 1901.

105 MoD, Whitehall Library, Confidential Telegrams nos. 558 and 559, p. 284, Secretary of State for War to Kitchener, 24 June 1901, and Kitchener to Secretary of State for War, 25 June 1901.

106 F. Pretorius, 'Confronted with the Facts: Why the Boer Delegates at Vereeniging Accepted a Humiliating Peace to End the South African War, 31 May 1902' in S. M. Miller (ed.), *Soldiers and Settlers in Africa, 1850–1918* (Leiden: Brill, 2009), pp. 195–217.

CONCLUSION

Railways and the preparation for war, 1914

In half a century the British army had demonstrated a sustained interest in the potential of railways for the conduct of their military activities. From the earliest interest in railways as an adjunct to their public-order duties through experimentation with newly laid lines in the Crimea and Abyssinia, to the exploitation of existing railroads in India and Egypt, the military recognized the transformative potential of being able to move men, arms, horses and supplies relatively quickly and of being able to deploy such forces in concentrated form and in good condition. The potential of the railway, a British invention, captured the imagination of many soldiers, not merely Royal Engineers but also senior commanders both at home and overseas. Doubtless the appeal had been enhanced by the expansion of the rail network at home and in the colonies, especially in India, and by the remarkable feats of engineering involved in bridge building, tunnel excavation, enhanced engine capacity and signalling *inter alia*. The massive scale of Victorian rail investment, passenger demand and pervasive imagery through rail art (lithographs, paintings and prints), photography and literary composition all ensured a prominence for the railway whether in commercial or military usage. Above all, railways were intimately associated with the imperial mission, penetrating the hinterland, connecting large landmasses, and enhancing both the security of the colonies and their prospects for commercial exploitation.

In his famous work *Small Wars: A Tactical Textbook for Imperial Soldiers*, Colonel Charles E. Callwell described colonial campaigns as fought 'rather against nature than against hostile armies'.[1] Constructing a railroad was only one option among several in overcoming difficult terrain; in making the choice much depended upon the proximity and movements of the enemy, climatic constraints, the availability of riverine alternatives, and factors of time, cost and the prevalence of disease locally. Evaluating these factors was far from an exact science except where the mission had to be undertaken rapidly and railways could not be built in time. As the great debates over the method of relieving Gordon in Khartoum demonstrated, political, bureaucratic and inter-service rivalries could influence the outcome. Choosing the railway option, too, had a symbolic significance, for if it was anything other than a short support line, as in Abyssinia, the railway embodied a permanent presence that could stretch beyond the immediate campaign and so had political implications from the outset.

Military interest, nonetheless, had focused initially upon the operational potential of the railway – its role in concentrating a military force whether in an internal or expeditionary role, reinforcing that body and sustaining its presence on active service. If railways were unlikely to be the sole source of transportation (and even in the Sudan the railway terminus remained at Fort Atbara), they enabled relatively large-scale forces to be deployed into a hinterland and to function effectively once they were deployed. In modern parlance railways were both 'force enablers' and 'force multipliers'. They also met another expectation of military operations, rooted in the public consciousness since the first winter in the Crimea, namely the prompt removal of the sick and wounded for medical treatment. This operational capacity had ensured that railways would be considered as instruments of military strategy, either responding to an invasion or deterring it by demonstrating the power to do so. The ability to move, concentrate and project military forces rapidly, whether domestically or on the north-west frontier, underpinned the thinking, which sought to exploit rail power in strategic missions. Conversely the precipitate use of railways could prove disastrous, as demonstrated by the first relief mission in China (June 1900).

The first relief mission to Peking (10–25 June 1900)

The rebellion of Boxers, an anti-Christian, anti-foreign movement in China, led to attacks on missionaries, the destruction of property belonging to foreign interests, and threats to the infrastructure laid by foreigners, including the Yangtsun bridge, a crucial rail link between Peking (Beijing) and Tientsin. By early June 1900 a sense of impending crisis gripped the legations in Peking, and although the formal siege of the foreign legation area had not yet begun, panicking diplomats pleaded for military relief from their armed forces near the coast. On 10 June 1900 Vice-Admiral Sir Edward Seymour responded by commandeering five trains and over 100 coaches to convey a relief expedition from Tientsin (see Map 7). He assembled a mixed force of some 2,100 men, 900 of whom were British, predominantly naval forces, and the remainder Russian, German, American, French, Japanese, Italian and Austrian troops. Most of the British travelled in the first

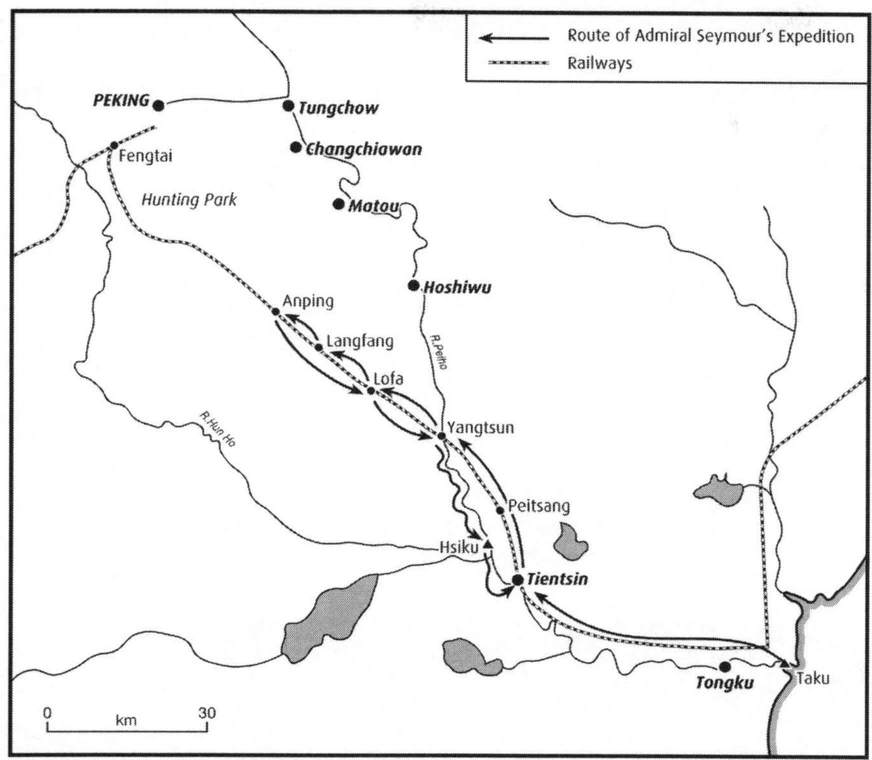

Map 7 Route of the first Peking relief expedition

train with a number of trucks carrying spare rails, sleepers and repair equipment, several English engineers and drivers, and seventy Chinese coolies. The remaining British and other troops travelled in the next three trains, while the final train carried supplies and was expected to keep the expedition sustained by shuttling back and forth to Tientsin. Sharing Seymour's confidence, Clive Bigham, a young diplomat who travelled with the expedition, envisaged the rail journey of some 129 km being completed in 'a few hours perhaps, at the most of a day, and then to an easy and successful march into the Tartar City, and the resumption of diplomacy'.[2]

The trains covered nearly half the distance on the first day before they encountered signs of Boxer rail sabotage at Lofa and had to halt to repair damaged track and a bridge. On the following day they reached Langfang, about 51.5 km from Peking, where the Boxers had not only torn up rails, burnt sleepers and damaged girders but had also destroyed the station watering tanks. As water had to be collected by pail for the engines and the men, and the repairs proved extensive, the pace became chronically slow. On the third day, the Boxers launched their first of a series of attacks on the trains. By 14 June communication with Tientsin was lost, as more Boxers had seized the bridge at Yangtsun and were destroying track to the rear as well as ahead of the expedition. By now the relief force was stretched to meet attacks from front and rear, but knew that the Peking legations were in a better shape (they would survive the ensuing siege for fifty-five days) while it could make scant progress: 'four miles [6.4 km] a day was all that could be done', recalled Bigham.[3] Nor was it possible to march across country from Langfang, as 'we were absolutely without transport, and directly in front of us lay the South Hunting Park, which was packed with the camps of the Peking Field Force'.[4]

On 17 June all the trains moved back towards Tientsin but, by 19 June, found the bridge at Yangstun wrecked and the station destroyed. As imperial cavalry had joined the Boxers in pursuit of the last train, all hope of further movement by railroad had evaporated (see Figure 14). Seymour and his men decided to retreat via the Peiho River; they

Figure 14 'The Expedition to Tientsin: Abandonment of Trains on June 19 near Yangtsun', *Illustrated London News*, vol. 117, no. 3202 (1 September 1900), p. 299

seized four junks in which they placed the wounded and their supplies and marched down the left bank, being harassed all the way by masses of imperial cavalry and horse artillery. Fortunately they found refuge in the well-stocked and fortified Hsiku Arsenal, which they held until they were relieved on 25 June by a Russian force with 500 British seamen under the command of Captain David Beatty. Seymour's force had lost 62 men and suffered 223 wounded.[5]

Undoubtedly Seymour had faced a real dilemma when he received the original intelligence, but his response, though gallant, was certainly imprudent. A rapid advance by the railway may have seemed an obvious option, but to make one without any means of reconnaissance at a time of considerable uncertainty, and across a countryside scarred by the ravages of the Boxer rebellion, was imprudent at least. 'Little did we think', wrote Bigham, 'that we should never get to Peking, and that when we struggled back to Tientsin with a seventh of our force killed and wounded, the station, the settlement, and the many signs of civilization that we now saw and took pride in would be burnt and desolated ruins, riddled with shot and shell and disfigured by rotting corpses.'[6] The second and successful relief expedition would eschew the railway and move up the Peiho River and across country to Peking.

Railway organization

Irrespective of whether the envisaged use of railways was operational and limited to the parameters of a specific campaign, or potentially strategic in defence of the homeland or India, army commanders understood that the railways in question had to be organized, managed and, if in hostile territory, protected. They had observed how railways could be used and misused in foreign wars, and Girouard had brought much of that understanding to bear in the Sudan and South Africa. In accounting for his approach, he referred to *Les chemins de fer pendant la guerre de 1870–71, par M. Jacqmin* as a demonstration of the way in which the railway arrangements of France in 1870 had descended into chaos by comparison with the 'stringent regulations' employed by the Germans. In time of war, reckoned Girouard, paramount authority had to reside with a director of railways and his associates in their particular stations or districts. 'Civil railway officials', he observed, 'have been heard to say that attacks by the enemy on the line are not nearly so disturbing to traffic as the arrival of a friendly general with his force.'[7]

In writing his history of the railways in the South African War, Girouard was not merely chronicling the achievements of the railway corps, including the additional services of '5,000 Europeans and 12,000 natives'.[8] As he explained in a post-war interview with Lord Esher, who was a confidant of the king and a member of the royal commission established to investigate the conduct of the war, lessons should be learned from the war, and a substantial railway staff should be created in peacetime. An intermediary staff, he affirmed, should be organized and trained in peacetime, charged with collating railway information, preparing manuals, organizing an imperial corps, all the details of home defence and the transport of forces in peacetime. In addition Girouard believed that a central railway authority should have three companies of railway technical soldiers at its disposal in peacetime.[9]

The proposal divided opinion among senior Royal Engineers. Kitchener endorsed Girouard's ideas, arguing that a military railway department should be formed 'whose duty it would be, not only to take over and work the railways during war, but also to collect all information regarding the railways in any possible theatre of operations'.[10] General Sir Richard Harrison, the inspector-general of fortifications, agreed that the railway companies should be larger and better trained in peacetime, with access to more stock, which could be used for peacetime training.[11] Conversely Major-General William Salmond, who had served as the deputy adjutant-general, RE, throughout the war, reckoned that it would be 'a pity to overload us in peace with a large railway service'. He questioned the veracity of Girouard's history, especially the emphasis upon improvisation from the outset. On the contrary Salmond claimed that he had sent Girouard a booklet entitled *Rules and Regulations for the Working of Railways in War in Foreign Countries* when he was appointed as director of railways in South Africa. He affirmed that this booklet, coupled with the railway companies in service, had provided a sufficient nucleus in peacetime, and opposed the whole concept of specialization in the Royal Engineers. He asserted that it would 'sacrifice your elasticity. What are Engineers but tradesmen, mechanics; good tradesmen all round?' Over-specialization, he claimed, would damage recruiting and inhibit the desire to serve overseas, where engineers 'can do all sorts of things'.[12]

However self-serving, such sentiments tapped into two strands of post-war thinking: first, a belief that the army had just prevailed in a protracted but profoundly abnormal war, and second, that although lessons had to be learned from the conflict, and changes in weapons and tactics had to follow, these reforms were unlikely to succeed if they presumed a significant increase in peacetime expenditure.[13] Just as the army had no intention of planning to engage in another costly, counter-guerrilla war over a vast territory, in which they would replicate their recent usage of railways and of mounted infantry, so there were cheaper options to consider. Salmond, like Major-General Sir Elliot Wood, formerly the Commanding Royal Engineer at Aldershot, saw salvation in the expertise of civilian railways and the readiness of civil engineers to join the reserves. As Salmond stated, 'I do not myself think that it is necessary to make a large peace railway organisation. I would look more, as I have said, to drawing in, in some way, aid from the commercial railway companies of the United Kingdom.'[14] A large railway department failed to evolve from the war, and Girouard served instead as the commissioner of railways in South Africa until the financial crash of 1904; this post was followed by a couple of staff appointments at home, and then two more substantive roles in Africa as high commissioner in Nigeria (1907–9) and later as the governor and commander-in-chief, East Africa Protectorate (1909–12).

Russo-Japanese War (1904–5)

Meanwhile the outbreak of war between the armed forces of the largest power in Europe and Japan, whose army and navy were at least perceived to possess the advantages of modern armaments and organization, ensured that this would become the most extensively observed war of the Edwardian era. Just as foreign journalists and military attachés rushed to observe the massive battles on land and sea, photographers and war artists found international audiences for their work. Sir Ian Hamilton was the first of British military attachés to arrive from India, and although others followed and reported from both sides, his reports, clearly sympathetic to the Japanese side, established him as the doyen of British observers. His two-volume work, *A Staff Officer's Scrap-Book during the Russo-Japanese War*, appeared shortly after the war and was reprinted in 1912. He recognized that railways would have a fundamental effect on the war, since when one army remained 'close to its base, the other hangs dependent at the extremity of a single line of railway thousands of miles long, like a soap-bubble at the end of a churchwarden'.[15] Yet Hamilton was much more interested in the tactical lessons of the war: the ineffectiveness of the cavalry; the role of artillery–infantry co-operation in sustaining offensives, pressed home at the point of a bayonet; the heavy casualties inflicted by modern weapons, including concentrated gunnery; and the tactics of envelopment.[16] Several official and semi-official publications,[17] coupled with an array of lectures and articles in military journals, amplified these tactical themes,[18] with some works alluding to the importance of railways as a means of reinforcement and resupply, especially on account of the prodigious expenditure of heavy ordnance.[19]

Lieutenant-Colonel Charles à Court Repington, the military correspondent of *The Times*, reckoned that the way in which Russia had managed the single-line Trans-Siberian Railway had a more immediate significance for the British empire. Although he had anticipated that Japan would win, he saw many lessons for Britain in a struggle between an island empire and a large, continental power. In a profusion of articles, later revised and updated with more accurate information in a single volume, *The War in the Far East*, Repington revealed a fascination with the capacity and management of the Trans-Siberian Railway as well as the local railways constructed in Manchuria, which linked Port Arthur, Kharbin and Vladivostok. By despatching twelve to fifteen trains a day along the latter route, Russia could meet a Japanese attack or move forces from one front of strategic deployment to another.[20] Repington was impressed, too, that the Trans-Siberian Railway had 'worked at full pressure', that the railway administration 'steadily improved' during the war and had sustained between '230,000 and 250,000 combatants' at the front. He noted that Russia had nearly doubled its power of reinforcement and supply 'in the course of the year . . . a lesson that we shall only neglect at the cost of the end of our rule in India'.[21] In a major chapter on 'Our Warning from Manchuria', he concluded that

> The warning of Manchuria lies in this pregnant fact, that it has given a practical illustration of the power of Russia to assemble and maintain a great army many thousands of miles from Western Russia by means of service of a single line of railway. The precedent of the Russian concentration in Manchuria must inevitably recur in every future plan for the attack or defence of India . . .[22]

Repington knew that the defence of India, and the issue of reinforcements in the event of war with Russia, had been investigated by the Committee of Imperial Defence (CID), an inter-departmental body which had been created in December 1902. So important was the issue that it had been raised as an issue for investigation at the first meeting of the CID on 18 December 1902,[23] and subsequent inquiries, expressed in a flurry of papers over the next three years, had revealed the immense complexity of the problems involved. On the basis of theoretical calculations about Russian rail-laying abilities and the carrying capacities of Russian railways, the War Office estimated that the Russians could deploy 150,000–200,000 men in the Afghan theatre by the end of the fifth month of war, and some 500,000 if the Tashkent–Orenburg railway was completed on schedule in 1905.[24] The number of forces, and more importantly of reinforcements that might be needed to defend India, depended upon the feasibility of deploying a forward defence along a line from Kabul to Kandahar. This would involve planning to build railroads across Afghanistan and possibly across Persia towards the vulnerable area of Seistan. Neither the CID nor Arthur J. Balfour, the prime minister, who took a keen interest in strategic issues, wanted to compromise the buffer status of Persia and Afghanistan between the British and Russian empires. The diplomatic concerns centred upon a Russo-Persian convention of 1900 that bound Persia not to construct railways or allow any foreign power to do so until 1910,[25] and a belief that an attempt to extort railway-building concessions from Amir Habiballah Khan would place his fidelity at risk. Britain's friendly relations with amir were seen to depend upon its respect for the independence of Afghanistan.[26]

Complicating these assumptions were the estimates from the political and military authorities in India. In a joint memorandum Lords Curzon and Kitchener challenged the British calculations of Russian railway carrying capacities as far too low, and confirmed the centrality of checking Russian ambitions towards Seistan, but indicated a reluctance to set precise numbers on the scale of reinforcements that India might require. Curzon assumed that India would get reinforcements of 100,000 even before the – Tashkent–Orenburg railway was completed,[27] and that it might need more. Kitchener subsequently reckoned that once the Tashkent–Orenburg railway was completed, and its Termez spur added, 135,614 officers and men might be needed as reinforcements. This estimate exceeded the assessment of the CID, which confirmed at its forty-seventh meeting (22 June 1904) that Britain could not provide reinforcements in excess of 100,000 men during the first twelve months of war.[28]

While Kitchener continued to dispute the adequacy of this calculation, Sir George Clarke, the secretary of the CID, emerged as the *bête noire* of the Indian authorities, challenging every aspect of the logistical and transport assumptions underpinning India's forward policy. He warned Balfour that concessions to Kitchener could make 'India the predominant partner in Imperial defence',[29] and that subsequent calculations by the general staff about Russia's proposed rate of railway construction overlooked the difficulty of 'railway making in a hill country occupied by well-armed tribesmen'.[30] He also applied a remorseless logic to another dimension of the controversy, namely the number of camels that would be required to support five Indian divisions as they advanced from their railheads and remained in Afghanistan for a year. If they undertook fifteen marches from their railheads, they might require 'considerably over 3,000,000 camels, or their equivalent, assuming no forage locally for transport animals'. While the numbers would vary if two divisions found forage in Kabul,

Clarke insisted that animal transport would still impose 'sharp limitations on the number of troops that can be maintained in Afghanistan'.[31]

Repington, having enjoyed 'a good wrangle with George Clarke and his merry men over the railway question in Central Asia', wrote to Kitchener's secretary, Raymond Marker, convinced that Clarke was underestimating the railway capacities of Tsarist Russia. He maintained that 'Russia can run 18 trains a day over the line to Kushk, and over its eventual extension, with ease, provided she takes note of her experiences in Manchuria.' He even envisaged that Russia could manage thirty trains a day on a single track 'with sidings 5 miles apart – 8 kilometres'. As a consequence, Repington estimated, there could be '500,000 Russians on the Helmund [*sic*: Helmand] or near it in nine months and nothing expected from the country'.[32] Underpinning this concern was a more profound anxiety; Repington was worried that Clarke could exercise undue influence over the incoming secretary of state for war, Richard B. Haldane.[33]

At the eighty-fifth meeting of the CID on 9 March 1906 Haldane asked the foreign secretary, Sir Edward Grey, how the outcome of the Russo-Japanese War and the Anglo-French entente had affected the probabilities of Russian aggression. Grey doubted that Russia, following its military and naval reverses in East Asia, and the revolution that followed the war, would be able to undertake any 'serious campaign against India' for 'a number of years'. He intimated, too, that the Russian government had expressed a wish to reopen negotiations with Britain in the hope of securing an agreement similar to that embodied in the *entente cordiale*.[34] As Grey was merely alluding to the possibility of a diplomatic solution to the question of Indian defence, the debates resumed at the eighty-eighth meeting of the CID. On this occasion Haldane, now in the process of devising his reforms of the home army, argued against giving India a definite commitment about the size of Indian reinforcements, and Grey endorsed this point of view. As John Morley, the secretary of state for India, was alarmed lest confirmation of this recommendation might prompt Kitchener to demand an increase in the size of the Indian garrison, Herbert Asquith, the prime minister, referred the issue to a sub-committee.[35]

Morley chaired the sub-committee, which submitted its report on 1 May 1907. After extensive hearings, they learnt that all the authorities, Kitchener included, no longer anticipated large-scale aggression or a direct assault on India by Russia. Both Kitchener and Sir William Nicholson, then quartermaster-general to the forces, envisaged the possibility of a more insidious policy, involving incremental movements into Herat and Afghan Turkestan, with an advance by stages on to Kandahar and Kabul over several years, possibly constructing a 'railway between Kabul, Ghazni and Kandahar'. Whatever the form of a threat, the sub-committee accepted that the British had to hold 'the Kabul–Kandahar alignment', and that this might require extending railways into the tribal regions, even if a 'further extension to Dakka' (in Afghanistan) 'must depend upon political circumstances'. Ultimately the sub-committee agreed that if India aimed to establish superiority over the Russians at decisive points, it would need reinforcements of 100,000 men during the first year of war. The CID accepted this recommendation at its ninety-eighth meeting on 30 May 1907,[36] and the issue was resolved eventually by the signing of the Anglo-Russian convention on 31 August 1907.[37]

Railways and 'WF' planning

Although the CID periodically returned to the issue of Indian reinforcements,[38] several departments in the War Office focused upon the railway requirements of a more immediate contingency. Ironically, another article by Repington in *The Times*, warning Germany not to expect British disinterest in the event of Franco-German war,[39] had triggered the Anglo-French discussions. It had prompted Major Victor J.-M. Huguet, the French military attaché and a friend of Repington, to inform the correspondent of the French government's anxieties about how Britain would act in the event of a war between France and Germany. After Repington passed this message on to Grey,[40] and the latter informed Haldane that in the event of a war between France and Germany, popular feeling might compel the government to ask him what support he could proffer,[41] the war minister met Grey at Berwick four days later. The two men agreed that secret discussions should be opened between the British and French general staffs. Two days later Haldane instructed Sir Neville Lyttelton, the chief of the general staff, that communications should begin between Sir James Grierson, the director of military operations (DMO), and Huguet in London. He also ordered a review of existing mobilization arrangements, which showed that Britain could deploy an army of only 80,000 in the field within a period of two months.[42]

As the *entente cordiale* agreed between Britain and France in 1904 had not involved any formal commitments between the two countries, the preliminary talks had to be fairly tentative and absolutely secret. After a couple of months the political stimulus faded as political détente followed the Algeciras conference of April 1906.[43] In retrospect the War Office admitted that at this time

> on our side no arrangements were made for moving the troops from their places of mobilization to the ports of embarkation, because the General Staff were expressly forbidden to take the railway companies into their confidence, and the Admiralty was only unofficially approached to provide the necessary sea transport . . .[44]

It was only in 1910, when Major-General Henry Wilson succeeded Grierson, as DMO that the work begun by Grierson and Huguet was reviewed seriously and definite steps taken to devise the plans for the movement of the British Expeditionary Force (BEF) overseas. Under Wilson's direction the whole war organization of the British army was subordinated gradually under his directorate, especially MO(1), headed by Colonel G. M. Harper, and thereafter the entire 'organization, mobilization and training in the direction of this special plan, known as the "W. F. Scheme" ["With France" Scheme]'.[45] The profoundly Francophile Wilson was incensed by the lack of drive and direction from General Sir William Nicholson, then chief of the imperial general staff. 'Nick's total lack of interest in our preparations for war', wrote Wilson, 'is deplorable.'[46] Having spent five months since August 1910 seeking information on the mobilization arrangements, he sought the support of Nicholson to 'make detailed arrangements for railing Expd. Force to point of embarkation. At present absolutely nothing exists which is scandalous.'[47] During a 'serious talk' with Nicholson on 10 January 1911, Wilson catalogued all the main deficiencies, including 'no train arrangements to ports', and, on the following day, he offered 'to take railway concentration & Admiralty arrangements off QMG [quartermaster-general] & do these myself'.[48] Wilson supplied Nicholson with a detailed memorandum, indicating the danger that existed because of the lack 'of those elemental measures necessary for the methodical mobilization, entrainment, & embarkation of the Expeditionary Force'.[49] Nicholson duly approached Haldane, who then apprised Grey, and the foreign secretary agreed that the military planners could now approach the railway companies. 'This is good', wrote Wilson in his diary of 20 January 1911.[50]

By March 1911 Wilson characterized his meetings with the chairman of London and South Western Railway, who would become the medium of communication with the other railway companies, as 'satisfactory', and claimed that the 'railway work progresses well'.[51] They agreed that Southampton would become the main port of embarkation for troops, and accepted that seventy trains was the maximum number that could be run into the port in every twenty-four-hour period. They resolved, too, that Avonmouth would be the main port of departure for military vehicles, and Newhaven for stores, and that every effort should be made to limit the number of main lines used to guard against sabotage. Commands made out all the tables for railway movement up to the completion of mobilization, including the trains for the collection and distribution of horses. MO(1) produced tables showing all units of the Expeditionary Force, sorted into train-loads, the day they were to move, and their destination. From this information the railways worked out the timings, with the date, place and hour of entrainment filled into each unit's mobilization scheme. The railway movements to the ports of embarkation were made out for each day of mobilization on 'Form 129/Mobilization/139' and were forwarded to the DMO by 19 May 1911.[52]

On 6 July 1911 Wilson was able to assure Nicholson that 'given the necessary information, the railway companies can rail the troops to the ports of embarkation and the Admiralty can ship them as fast as required for the particular campaign we have in view'. Serious problems persisted in the shortage of officers and horses for mobilization, the lack of certain war establishments, the quality of certain arms and ammunition, and the impact of any accelerated mobilization plan.[53] Completing the residual mobilization arrangements would take many months, and although Harper advised Wilson in April 1912 that the 'necessary railway arrangements for the movement of troops to the ports of embarkation have been worked out', it was not until November 1912 that complete railway timetables were sent to the commands.[54] The little pamphlet, 'Instructions for Entrainment and Embarkation', was also produced.[55] All units would know the time of their rail departure on the specific day of mobilization, their train numbers and their scheduled arrival time at Southampton. These rail moves due to begin on the second day were scheduled to be largely completed by the fifteenth day. At Southampton the duties of personnel were described in detail, with an embarkation commandant, assisted by a staff of 162 men, who could co-ordinate movements into the dock area, allocate troops to ships, and supervise the loading process. To avoid confusion every trainload of troops was treated as a self-contained unit and loaded as such, even if it meant that some vessels sailed with spare capacity. Of all aspects of the mobilization arrangements, the railway arrangements received particular care and precise calculation, ensuring that the units of the BEF, whether based in mainland Britain or Ireland, could move smoothly to their ports of embarkation. As Samuel Williamson remarked, 'No detail, no item was left to chance or doubt.'[56]

Matching British planning to French facilities and transport capacities added a further layer of complication. The three Channel ports in France selected for the reception of British troopships were Le Havre, which was able to receive thirty ships per day, Rouen, twenty ships per day, and Boulogne, eleven ships per day. Only sixty trains per day could leave these ports: twenty-five trains daily from Le Havre, fifteen daily from Rouen and twenty daily from Boulogne. As a French trainload could carry the equivalent of two British trainloads, this meant that all units had to be reallocated on arrival and that component parts of a French trainload might arrive on more than one ship. Accordingly, each British battalion had to receive two serial numbers, one for his movement on British railways and ports and the other for French railways. All these arrangements were then examined in a staff ride held at Amiens during the latter part of June 1914, where the railway planning was tested thoroughly in the presence of officers from MO(1). This was not simply a review of the disembarkation, train loading and rail movements but also how one of alternative routes could cope if accidents occurred, such as tunnels being blown in or trains becoming derailed. General Sir Percy Radcliffe, who worked in MO(1) from September 1913, reckoned that the 'elasticity and adaptability' of the planning were among its most important features. It allowed the planners to respond to unforeseen circumstances, which happened in August 1914 when the despatch of the BEF was delayed, two of the six divisions were withheld initially, and then the French base had to be changed after the retreat from Mons.[57]

Conclusion

Over seventy years the British army's interest in railways had come full circle. Just as it had perceived the railway as a potentially key enabling instrument when required to respond to certain instances of civil disorder in the Chartist era, so the army recognized that the railway was a prerequisite if it intended to deploy and concentrate the BEF on the Belgian border by the twelfth day of mobilization. The learning process had begun by utilizing the assistance of civilian railway engineers in the Crimean War and had expanded gradually by observing foreign wars, experimenting in colonial contexts and developing construction, operational and railway management expertise among the Royal Engineers. The railway building programmes on the north-west frontier and across the Nubian Desert testified to the value placed on railways by government officials and senior officers, and to the managerial skills, courage, discipline and resilience of those involved. The South African War then tested those qualities in the most demanding of circumstances, not merely the sustained attacks upon the railway by enemy forces but also the operation of railways across a vast and often inhospitable landmass, a theatre far larger than anything that would be encountered in the First World War.

The planning requirements for entry into a European war posed challenges of mobilization – of reservists and Territorials returning to their units, and then of regular forces moving to ports of embarkation – and these challenges were unprecedented in their scale and rapidity. Radcliffe recalled that

> When things began to look ugly in the Summer of 1899 the War Office asked the Admiralty how long it would take to provide transport for 49,000 men and 7,900 horses. The answer was 4 to 5 weeks. The gap to be bridged between this and the completion of concentration on the Belgian frontier, by the 12th day, of 165,894 men and 60,368 horses was thus a pretty big one.[58]

Such matters could be addressed in professional and thoughtful manner only by engaging with civilian and naval expertise, and by collaborating, albeit in a covert manner, with a potential ally in a totally new operational theatre. Had the army not appreciated the military potential of railways in imperial wars, it would have lacked the expertise to exploit it as effectively as it did in August 1914.

Notes

1 Callwell, *Small Wars*, p. 57.
2 C. Bigham, *A Year in China 1899–1900* (London: Macmillan, 1901), p. 170; for good general accounts of the relief mission see D. Preston, *The Boxer Rebellion: The Dramatic Story of China's War on Foreigners that Shook the World in the Summer of 1900* (New York: Berkley Books, 1999), pp. 90–104; P. Fleming, *The Siege at Peking* (Edinburgh: Birlinn, 1959), ch. 6.
3 Bigham, *A Year in China*, p. 175.
4 Ibid. pp. 176–7.
5 Preston, *Boxer Rebellion*, p. 104.
6 Bigham, *A Year in China*, p. 170.
7 TNA, WO 108/378, Girouard, *History of the Railways*, p. 11.
8 TNA, CAB 17/11, Committee of Imperial Defence, 'Railway Organisation in Theatres of War, 1903'.
9 Ibid.
10 PP, *Report of Her Majesty's Commissioners Appointed to Inquire into Military Preparations and other Matters Connected with the War in South Africa*, Cd 1789 (1904), XL (Elgin Report), p. 110, and Lord Kitchener (Q. 219) in evidence before the Elgin Commission, Cd 1789, vol. 1 (1904), XL, p. 12.
11 PP, Gen. Sir R. Harrison (Q. 2052), in evidence before the Elgin Commission, Cd 1789, vol. 1 (1904), XL, p. 96.
12 PP, Maj.-Gen. W. Salmond (Qs. 2119 and 2121), in evidence before the Elgin Commission, Cd 1789, vol. 1 (1914), XL, p. 99.
13 E. M. Spiers, 'Between the South African War and the First World War, 1902–14' in H. Strachan (ed.), *Big Wars and Small Wars: The British Army and the Lessons of War in the Twentieth Century* (London: Routledge, 2006), pp. 21–35, at pp. 21–2, 25.
14 PP, Salmond (Q. 2119) and Maj-Gen. Sir E. Wood (Q. 2275), in evidence before the Elgin Commission, Cd 1789, vol. 1 (1904), XL, pp. 99 and 100.
15 Lt.-Gen. Sir I. Hamilton, *A Staff Officer's Scrap-Book during the Russo-Japanese War*, 2 vols. (London: Edward Arnold, 1905), vol. 1, p. 11; see also Wolmar, *Engines*

of War, pp. 122–3, and F. Patrikeeff and H. Shukman, *Railways and the Russo-Japanese War* (London: Routledge, 2007).

16 Hamilton, *Staff Officer's Scrap-Book*, vol. 1, pp. 131–2, 238, 324–5; vol. 2, pp. 98–9, 261.

17 Historical Section, Committee of Imperial Defence, *Official History of the Russo-Japanese War*, 3 vols. (London, 1908–9); War Office, *The Russo-Japanese War: Reports from British Officers* (London, 1908).

18 There is a useful list of these articles and lectures in K. Neilson, '"That Dangerous and Difficult Enterprise": British Military Thinking and the Russo-Japanese War', *War & Society*, vol. 9, no. 2 (1991), pp. 17–37.

19 Lt. C. R. Satterthwaite, 'Some Notes on the Action of the Japanese First Army at Liao-Yang', *RE Journal*, vol. 17 (1913), pp. 343–54, and Col. F. D. V. Wing, 'The Distribution and Supply of Ammunition on the Battlefield', *Journal of the Royal United Service Institution*, vol. 52 (1908), pp. 895–924.

20 The Military Correspondent of *The Times* [Lt.-Col. C. à Court Repington], *The War in the Far East* (London: John Murray, 1905), pp. 40–1.

21 Ibid. pp. 108, 355, 453 and 493–4.

22 Ibid. p. 597. The full extent of Repington's post-war revisions, particularly of the numbers engaged at the front, can be gauged by comparing ch. 49 with the original article 'Our Warning from Manchuria', *The Times*, 6 December 1904, p. 4.

23 TNA, CAB 2/1, first meeting of Committee of Imperial Defence CID, 18 December 1902: for a full account of the debates, see J. Gooch, *The Plans of War: The General Staff and British Military Strategy c. 1900–1916* (London: Routledge & Kegan Paul, 1974), pp. 198–237.

24 TNA, CAB 6/1/6D, 'Memorandum on the Defence of India', 10 March 1903, pp. 5–6.

25 Ibid. 'Second Instalment of Draft Conclusions on Indian Defence by Mr. Balfour, Dealing Chiefly with Seistan', 20 May 1903.

26 BL, Add. MSS 49720, fo. 266, Balfour MSS, A. J. Balfour to W. St J. Brodrick, 17 December 1903.

27 TNA, CAB 6/1/30D, 'Memorandum by the Viceroy and the Commander-in-Chief on the Provisional Report of the Defence Committee on Indian Defence', 3 July 1903, and CAB 6/1/37D, 'Despatch of Reinforcement and Supplies to India in the Event of War', Viceroy to Secretary of State for India, 8 January 1904.

28 TNA, CAB 2/1, forty-seventh meeting of the CID, 22 June 1904; CAB 6/2, 'Demands for Reinforcements by the Government of India', 20 February 1905.

29 BL, Add. MSS 50836, fo. 116, Sydenham MSS, 'Note on the Discussion of the 22nd Inst. Indian Reinforcements, Sent to A. J. B. 24.11.1904'.

30 BL, Add. MSS 49701, fo. 28, Balfour MSS, Sir G. S. Clarke to Balfour, 18 January 1905.

31 TNA, PRO 30/57/34, Kitchener MSS, Clarke to Kitchener, 7 July 1905.

32 BL, Add. MSS 52277B, fos. 25–8, Kitchener–Marker MSS, Repington to R. Marker, 15 December 1905.

33 Ibid.

34 TNA, CAB 2/2, eighty-fifth meeting of the CID, 9 March 1906.

35 TNA, CAB 2/2, eighty-eighth meeting of the CID, 25 May 1906.

36 TNA, CAB 2/2, ninety-eighth meeting of CID, 30 May 1907; CAB 16/2, 'Report and Minutes of Evidence of a Sub-Committee of the Committee of Imperial Defence, Appointed by the Prime Minister to Consider the Military Requirements of the Empire as Affected by the Defence of India', 1 May 1907.

37 K. Neilson, *Britain and the Last Tsar: British Policy and Russia 1894–1917* (Oxford: Clarendon Press, 1995), p. 136; Gooch, *Plans of War*, p. 231.

38 TNA, CAB 4/3/116B, 'Report of the Standing Sub-Committee of the Committee of Imperial Defence Appointed to Enquire into the Question of the Oversea Transport of Reinforcements in Time of War', 16 June 1910.

39 *The Times*, 27 December 1905; see also C. à Court Repington, *Vestigia: Reminiscences of Peace and War* (London: Constable, 1919), pp. 262–3.

40 TNA, FO 800/110, Grey MSS, Repington to Sir E. Grey, 29 December 1905; see also A. J. A. Morris, *The Letters of Lieutenant-Colonel Charles à Court Repington CMG: Military Correspondent of The Times, 1903–1918* (Stroud, Gloucestershire: Sutton Publishing for the Army Records Society, 1999), pp. 22, 71–2.
41 National Library of Scotland, MS 5907, fo. 10, Haldane MSS, Sir E. Grey to Haldane, 8 January 1906.
42 Lord Haldane, *Before the War* (London: Cassell, 1920), pp. 30–2.
43 Gen. Sir P. Radcliffe, '"With France": The "WF" Plan & the Genesis of the Western Front', *Stand To!*, vol. 11 (1984), pp. 6–12.
44 TNA, WO 106/49/WH/1, 'Memorandum by Major Gorton'.
45 TNA, WO 106/49A/1, Maj.-Gen. Sir P. P. de B. Radcliffe, 'History of WF Scheme 1914'.
46 Imperial War Museum (IWM), Wilson diaries, H. Wilson, diary, 2 November 1910; for a full account of Wilson's role, see K. Jeffery, *Field Marshal Sir Henry Wilson: A Political Soldier* (Oxford: Oxford University Press, 2006), pp. 85–106.
47 IWM, Wilson diaries, Wilson, diary, 9 January 1911.
48 Ibid. 10 January 1911 and 11 January 1911.
49 TNA, WO 106/49/WH/1, 'Memorandum by Major Gorton on the Progress of the Scheme for the Despatch of the Expeditionary Force during the Half-Year Ending June 30 1911'.
50 Ibid.; IWM, Wilson diaries, Wilson, diary, 20 January 1911.
51 Ibid.13 and 17 March 1911.
52 TNA, WO 106/49/WH1, 'Memorandum by Major Gorton'; WO 106/49A/1, Radcliffe, 'History of WF Scheme 1914'; and P. T. Scott (ed.), 'The View from the War Office: A Preliminary Extract from the Previously Unpublished War Diary of General Sir Charles Deedes, KCB, CMG, DSO', *Stand To!*, vol. 11 (1984), pp. 14–16, at p. 15.
53 TNA, WO 106/49/WR/4(14), Wilson to Nicholson, 6 July 1911.
54 TNA, WO 106/50, Col. G. M. Harper to DMO, 17 April 1912, and IWM, Wilson diaries, Wilson, diary, 14 November 1912.
55 TNA, WO 106/49A/1, Radcliffe, 'History of WF Scheme 1914'.
56 S. R. Williamson, Jr, *The Politics of Grand Strategy: Britain and France Prepare for War, 1904–1914* (Cambridge, Massachusetts: Harvard University Press, 1969), p. 112; all railway timetables can be found in TNA, WO 33/665.
57 TNA, WO 106/49A/1, Radcliffe, 'History of WF Scheme 1914'.
58 Ibid. and Radcliffe, '"With France"', p. 8.

APPENDIX 1

Working methods on the Sudan railway, 1884–85

The system of plate-laying adopted was roughly as follows –

As soon as the material train, carrying about a quarter of a mile of permanent way, arrived, it was unloaded and the material packed on trollies by a party of Egyptians, 60 to 90 sleepers to a trolly [*sic*] and 30 rails to a pair. . . . The rails and sleepers were carried from them by parties of Egyptian soldiers – six men to a rail and one to a sleeper – and the latter thrown down roughly in position and then arranged by a sapper and four soldiers; the rails were placed upon them and the small material distributed by two or more men. Directly a trolly was emptied it was run back, being of course taken off the line if it met one with material.

The rails were placed in position, fish-plates put on, and bolts hand-tightened by a party of six Indians, accompanied by a man with a square who looked to the squareness of the joints. They were followed by six spanner-men who tightened the joints, and then came six sleeper squarer's, two of whom marked with chalk upon the rails the position of each sleeper. Immediately in front of the material trollies were six pairs of spikers for the joints; round curves two extra men were put on for the centres.

Behind the material trollies were four more spanner-men, who finished up tightening the joints, and then 30 pairs of spikers. Each pair of spikers full spiked a rail and then moved on to the corresponding rail fifteen lengths ahead. They were followed by four rough straighteners and the rough packers, about 30 men, who packed the line sufficiently to allow of the material train passing over it. Then came another straightening party of four.

The full packing party of about 40 men followed about half-a-mile in rear; behind them were the final straighteners – four men – and some 80 Egyptians boxing up. There was also a jim-crow party of four men to take out bad joints.

Source: Lt. M. Nathan, 'The Sudan Military Railway', p. 41.

APPENDIX 2

Indian volunteer railway units, 1869–1901

Year of formation	Name	Comments
1869	East Indian Railway Volunteer Rifle Corps	On 20 May 1890, it absorbed Sibpore College Volunteer Rifle Corps
1873	Eastern Bengal State Railway Volunteer Rifle Corps	
1879	Northern Bengal State Railway Volunteer Rifle Corps	In 1882, the two Bengal Railway corps were amalgamated as the Eastern Bengal Railway Volunteer Rifle Corps
1875	Great Indian Peninsula Railway Volunteer Corps	In 1902, it was absorbed as the 2nd Battalion, Midland Railway Volunteer Corps
1877	Bombay, Baroda and Central India Railway Volunteer Corps	In 1886, it absorbed the Ghadeshi Volunteer Rifle Corps, and in 1887 it was amalgamated with the Rajputana-Malva Volunteer Rifle Corps
1879	Tirhut State Railway Volunteer Rifle Corps	On 17 June 1892, it absorbed the Ghazipur Volunteer Rifle Corps
1884	Southern Indian Railways Volunteer Rifle Corps	
1885	Madras Railway Volunteer Corps	

1886	Sind, Punjab and Indus Valley Railways Rifle Corps	In 1888, the name was changed to 3rd Punjab (North-Western Railway) Volunteer Rifle Corps, and in 1892, to North-Western Railway Volunteer Rifles
1886	Southern Maratha Railway Rifle Corps	In1898, it absorbed the Bellary Volunteer Rifle Corps
1888	Bengal Nagpur Railway Volunteer Rifle Corps	In 1898, it absorbed the Orissa Volunteer Rifles, and in 1903, it was organized in two battalions
1890	Midland Railway Volunteer Corps	
1901	Assam Bengal Railway Volunteer Rifles	

Adapted from Gaylor, *Sons of John Company*, pp. 39–43; Roy, 'India' in Beckett (ed.), *Citizen Soldiers*, pp. 115–16.

SELECT BIBLIOGRAPHY

Manuscript sources

Blackburn Library
Tiplady diary

British Library (BL)
Balfour MSS
Broughton diaries
Dilke MSS
Gordon MSS
Kitchener–Marker MSS
Lansdowne MSS
Napier MSS
Peel MSS
Ripon MSS
Rose MSS
Sydenham MSS

Durham County Record Office (DRO)
Londonderry MSS

Guards Museum
Jeffreys diary

Hatfield House Archives (HHA)
3rd Marquess of Salisbury MSS

Imperial War Museum (IWM)
Wilson diaries

India Office Records (IOR), British Library
North Western Province Proceedings
White MSS

Liddell Hart Centre for Military Archives, King's College London
Maurice MSS

Ministry of Defence (MoD), Whitehall Library
Confidential Telegrams

Museum of Lincolnshire Life (MLL)
Fitzgibbon Cox diary

National Army Museum (NAM)
Burn Murdoch MSS
Churcher TS diary
Kitchener–Wood MSS
Meiklejohn, 'The Nile Campaign'
Roberts MSS

National Library of Scotland
Haldane MSS

Queen Mary, University of London Archives
Lyttelton Family Papers

Royal Archives (RA)
Duke of Cambridge MSS

Sudan Archive Durham University (SAD)
Farley memoir
Hunter MSS
Longe MSS (Creagh-Osborne diary)
Wingate diary

The National Archives (TNA)

Cabinet
CAB 4/3
CAB 17/11
CAB 37/12

Colonial Office
CO 96/103
CO 417/362

Committee of Imperial Defence
CAB 2/1 and 2/2
CAB 6/1, 6/2 and 6/3
CAB 16/2

Foreign Office
FO 65/1099
FO 78/4893
FO 78/4895
FO 78/5049
FO 800/110, Grey MSS

Home Office
HO 41
HO 45
HO 50

PRO 30/40, Ardagh MSS

PRO 30/57, Kitchener MSS

War Office
WO 6/78
WO 28/245
WO 30
WO 32
WO 33
WO 33/2A
WO 33/2B
WO 44
WO 80, Murray MSS
WO 106/56
WO 108/378, Lt.-Col. Sir E. P. C. Girouard, *History of the Railways during the War in South Africa, 1899–1902* (London: HMSO., 1903)
WO 130/378
WO 147

University of Leeds, Liddle Collection
Mann MSS

Parliamentary Papers (PP)

Army. Proclamations Issued by Field Marshal Lord Roberts in South Africa, Cd 426 (1900), LVI

Correspondence with Reference to the Proposed Construction of a CHANNEL TUNNEL, C. 3358 (1882), LIII

Dispatch Written by the Lord Lieutenant of Ireland to the Secretary of State, 2nd May 1839, Stating what Regiments could be Spared from Ireland, No. 30 (1844), XXXIII

East India (Frontier Railways, etc.), 223-Sess. 2 (1880), LII

Egypt No. 2 (1885): Correspondence Respecting British Military Operations in the Soudan, C. 4280 (1885), LXXXIX

Fifth Report from the Select Committee on Railways: 1844 (1844), XL

Railway (Balaklava to Sebastopol) (1854–55), XXXII

Report from the Joint Select Committee of the House of Lords and the House of Commons on the Channel Tunnel; together with the Proceedings of the Committee, Minutes of Evidence, and Appendix, C. 248 (1883), XII

Report from the Select Committee on Commissariat and Transport Services (Egyptian Campaign) together with the Proceedings of the Committee, Minutes of Evidence and Appendix, C. 285 (1884), X

Report from the Select Committee on the Abyssinian War: Together with the Proceedings of the Committee, Minutes of Evidence, and Appendix, no. 380 (1868–69), VI

Report of a Committee Appointed by the Secretary of State for War to enquire into the Administration of the Transport and Supply Departments of the Army, C. 3848 (1867), XV

Report of Her Majesty's Commissioners Appointed to Inquire into Military Preparations and Other Matters Connected with the War in South Africa, Cd 1789 (1904), XL (Elgin Report)

Report of the Gauge Commissioners, no. 34 (1846), XVI

Second Report from the Select Committee on Army and Ordnance Expenditure, no. 499 (1849), IX

Second Report of the Select Committee on Police (1852–53), XXXVI

Suakim–Berber Railway: Agreement between the Secretary of State for War and Messieurs Lucas and Aird for the Construction of a Line of Railway from Suakim towards Berber, C. 4325 (1884–85), XLVI

Suakim (Cost of Military Expedition), C. 360 (1884–85), XLVI

Printed primary sources (letters, diaries, memoirs, campaign histories and contemporary first-hand accounts)

Adaptation of Railways for Military Transport: Progress Report, 1877 (Simla: Government Press, 1877)

Anderson, J., *System of National Defence* (Edinburgh, 1853)

Andrew, W. P., *The Indus and its Provinces, their Political and Commercial Importance Considered in Connexion with Improved Means of Communication* (London: W. H. Allen & Co., 1858)

'An Officer', *Sudan Campaign 1896–1899* (London: Chapman & Hall, 1899)

Aspinall, A. (ed.), *Three Early Nineteenth Century Diaries* (London: Williams and Norgate, 1952)

Austin, Brig.-Gen. H. H., 'The Kabul River Survey', *RE Journal*, vol. 41 (1927), pp. 414–26

Beckett, I. F. W. (ed.), *Wolseley and Ashanti: The Asante War Journal and Correspondence of Major General Sir Garnet Wolseley 1873–1874*

(Stroud, Gloucestershire: The History Press for the Army Records Society, 2009)

Bennett, Lt.-Col. I. H. W., *Eyewitness in Zululand: The Campaign Reminiscences of Colonel W. A. Dunne, CB South Africa, 1877–1881* (London: Greenhill Books, 1989)

Bigham, C., *A Year in China 1899–1900* (London: Macmillan, 1901)

Birdwood, Field-Marshal Lord, *Khaki and Gown: An Autobiography* (London: Ward, Lock & Co., 1941)

Blake, Col. J. Y. F., *A West Pointer with the Boers* (Boston: Angel Guardian Press, 1903)

Bonham Carter, V. (ed.), *Surgeon in the Crimea: The Experiences of George Lawson Recorded in Letters to his Family 1854–1855* (London: Constable, 1968)

Borrett, Lt. H. C., *My Dear Annie: The Letters of Lieutenant Herbert Charles Borrett The King's Own Royal Regiment Written to his Wife, Annie, during the Abyssinian Campaign of 1868* (Lancaster; King's Own Royal Regiment Museum, 2003)

Burleigh, B., *Sirdar and Khalifa or the Reconquest of the Soudan 1898* (London: Chapman & Hall, 1898)

Childers, S., *The Life and Correspondence of the Right Hon. Hugh C. E. Childers, 1827–1896*, 2 vols. (London: John Murray, 1901)

Churchill, W. S., *The River War: The Sudan, 1898* (London: Eyre & Spottiswoode, 1899, reprinted London: Sceptic, 1987)

——, *London to Ladysmith via Pretoria* (London: Longmans, Green & Co., 1900)

——, *My Early Life: A Roving Commission* (London: Odhams Press, 1930)

Colomb, Vice-Admiral P. H., *Memoirs of Admiral the Right Honble. Sir Astley Cooper Key* (London: Methuen, 1898)

Colville, Col. H. E., *History of the Sudan Campaign, Compiled in the Intelligence Division of the War Office*, 3 vols. (London: HMSO, 1889)

Cromer, Earl of, *Modern Egypt*, 2 vols. (London: Macmillan, 1908)

De Wet, C. R., *Three Years War (October 1899 – June 1902)* (London: Constable, 1903)

Douglas, Bart., Sir G., and Sir G. D. Ramsay (eds.), *The Panmure Papers: Being A Selection from the Correspondence of Fox Maule, Second Baron Panmure, Afterwards Eleventh Earl of Dalhousie, K.T., G.C.B.*, 2 vols. (London: Hodder & Stoughton, 1908)

Driver, Nurse K., *Experience of a Siege: A Nurse Looks Back on Ladysmith*, Diary the Siege of Ladysmith, 6, revised edition (Ladysmith: Ladysmith Historical Society, 1994)

Elsmie, G. R. (ed.), *Field-Marshal Sir Donald Stewart: An Account of his Life, Mainly in his Own Words* (London: John Murray, 1903)

Fitzherbert, C., *Henry Clifford V.C. His Letters and Sketches from the Crimea* (London: Michael Joseph, 1956)

Haldane, Lord, *Before the War* (London: Cassell, 1920)

Hamilton, Lt.- Gen. Sir I., *A Staff Officer's Scrap-Book during the Russo-Japanese War*, 2 vols. (London: Edward Arnold, 1905)

Hart-McHarg, W., *From Quebec to Pretoria: With the Royal Canadian Regiment* (Toronto: William Briggs, 1902)

Henty, G. A., *The March to Magdala* (London, 1868, reprinted London: Robinson Books, 2002)

Holland, B. (ed.), *The Life of Spencer Compton Eighth Duke of Devonshire*, 2 vols. (London: Longmans Green and Co., 1911)

Holland, Maj. T. J., and Capt. H. Hozier, *Record of the Expedition to Abyssinia, Compiled by Order of the Secretary of State for War*, 2 vols. (London: Her Majesty's Stationery Office, 1870)

Hutchinson, Col. H. D., *The Campaign in Tirah 1897–1898: An Account of the Expedition against the Orakzais and Afridis under General Sir William Lockhart. Based (by Permission) on Letters Contributed to 'The Times'* (London: Macmillan, 1898)

Kaye, J. W., *Lives of Indian Officers*, 3 vols. (London: W. H. Allen, 1867)

Knight, E. F., *Letters from the Sudan* (London: Macmillan & Co., 1897)

Laband, J. C. (ed.), *Lord Chelmsford's Zululand Campaign 1878–1879* (Stroud, Gloucestershire: Alan Sutton Publishing for the Army Records Society, 1994)

Macdonald, J. H. A., *Fifty Years Of It: The Experiences and Struggles of a Volunteer of 1859* (Edinburgh: William Blackwood, 1909)

McFall, Capt. C., *With the Zhob Field Force 1890* (London: William Heinemann, 1895)

Malins, W., *A Plan for Additional National Defences* (London: J. Ridgway, 1848)

Martineau, J., *The Life of Henry Pelham Fifth Duke of Newcastle 1811–1864* (London: John Murray, 1908)

Massie, A., *The National Army Museum Book of the Crimean War: The Untold Stories* (London: Sidgwick & Jackson, 2004)

Maurice, Maj.-Gen. Sir F. B., and M. H. Grant, *History of the War in South Africa 1899–1902*, 4 vols. (London: Hurst & Blackett, 1906–10)

Maurice, Col. J. F., *Military History of the Campaign of 1882 in Egypt* (London: HMSO, 1887)

The Military Correspondent of *The Times* [Lt.-Col. C. à Court Repington], *The War in the Far East* (London: John Murray, 1905)

Morris, A. J. A., *The Letters of Lieutenant-Colonel Charles à Court Repington CMG: Military Correspondent of The Times, 1903–1918* (Stroud, Gloucestershire: Sutton Publishing for the Army Records Society, 1999)

Napier, Lt.-Col. the Hon. H. D. (ed.), *Letters of Field-Marshal Lord Napier of Magdala concerning Abyssinia, Egypt, India, South Africa etc.* (London: Simpkin Marshall Ltd., 1936)

Napier, Lt.-Gen. Sir W., *The Life and Opinions of General Charles James Napier, G.C.B.*, 4 vols. (London: John Murray, 1857)

Parker, C. S. (ed.), *Life and Letters of Sir James Graham Second Baronet of Netherby, P.C., G.C.B. 1792–1861* (London: John Murray, 1907)

Preston, A. (ed.), *In Relief of Gordon: Lord Wolseley's Campaign Journal of the Khartoum Relief Expedition 1884–1885* (London: Hutchinson, 1967)

Robins, Maj. C. (ed.), *The Murder of a Regiment: Winter Sketches from the Crimea 1854–1855 by An Officer of the 46th Foot (South Devonshire Regt) Annotated, and with Additional Material* (Bowdon: Withycut House, 1994)

—— (ed.), *Romaine's Crimean War: The Letters and Journal of William Govett Romaine Deputy Judge-Advocate to the Army of the East 1854–6* (Stroud, Gloucestershire: Alan Sutton for the Army Records Society, 2005)

Robson, B. (ed.), *Roberts in India: The Military Papers of Field Marshal Lord Roberts 1876–1893* (Stroud, Gloucestershire: Alan Sutton for the Army Records Society, 1993)

Russell, W. H., *The War: From the Landing at Gallipoli to the Death of Lord Raglan*, 2 vols. (London: George Routledge & Co., 1855)

Schikkerling, R. W., *Commando Courageous (A Boer's Diary)* (Johannesburg: Hugh Keartland, 1964)

School of Military Engineering, Chatham, *Instruction in Military Engineering*, vol. 1 (part V) (London: HMSO, 1878)

Scott, P. T. (ed.), 'The View from the War Office: A Preliminary Extract from the Previously Unpublished War Diary of General Sir Charles Deedes, KCB, CMG, DSO', *Stand To!*, vol. 11 (1984), pp. 14–16

Settar, S., and B. Misra (eds.), *Railway Construction in India: Select Documents*, 2 vols. (New Delhi: Northern Book Centre/Indian Council of Historical Research, 1999)

Spiers, E. M. (ed.), *Letters from Ladysmith: Eyewitness Accounts of the South African War* (Barnsley: Pen & Sword, 2010)

—— (ed.), *Letters from Kimberley: Eyewitness Accounts from the South African War* (Barnsley: Pen & Sword, 2013)

Stanley, H. M., *Coomassie and Magdala: The Story of Two British Campaigns in Africa* (London: Sampson Low, Marston Low & Searle, 1874)

Steevens, G. W., *With Kitchener to Khartum* (Edinburgh: William Blackwood & Son, 1898)

Stephenson, Sir F. C. A., *At Home and on the Battlefield: Letters from the Crimea, China and Egypt, 1854–1888* (London: John Murray, 1915)

Sterling, Lt.-Col. A., *The Highland Brigade in the Crimea: Founded on Letters Written during the Years 1854, 1855, and 1856* (Minneapolis: Absinthe Press, 1995)

Stewart, J. W., 'A Subaltern in the Sudan, 1898', *The Stewarts*, vol. 17, no. 4 (1987), pp. 223–8

Sydenham of Combe, Col. Lord, *My Working Life* (London: John Murray, 1927)

Todd, P., and D. Fordham, *Private Tucker's Boer War Diary* (London: Elm Tree Books, 1980)

Vetch, Col. R. H., *Life, Letters, and Diaries of Lieut.-General Sir Gerald Graham, V.C., C.B., R.E.: With Portraits, Plans, and his Principal Despatches* (Edinburgh: Blackwood, 1901)

——, *Life of Lieut.-General the Hon. Sir Andrew Clarke, G.C.M.G., C.B., CLE* (London: John Murray, 1905)

Viljoen, Gen. B., *My Reminiscences of the Anglo-Boer War* (London: Hood, Douglas & Howard, 1903)

The War Correspondence of The Daily News 1870 Edited, with Notes and Comments Forming a Continuous Narrative of the War between Germany and France1870, 2 vols. (London: Macmillan, 1871)

Ward, S. G. P. (ed.), 'The Scots Guards in Egypt, 1882: The Letters of Lieutenant C. B. Balfour', *Journal of the Society for Army Historical Research*, vol. 51 (1973), pp. 80–104

Wessels, A. (ed.), *Lord Roberts and the War in South Africa 1899–1902* (Stroud, Gloucestershire: Alan Sutton Publishing for the Army Records Society, 2000)

—— (ed.), *Lord Kitchener and the War in South Africa 1899–1902* (Stroud, Gloucestershire: Alan Sutton Publishing for the Army Records Society, 2006)

Wolseley, Col. Sir G., *The Use of Railroads in War: A Lecture Delivered at Aldershot, on the 20th January 1873* (London: Edwin S. Boot, 1873)

——, *The Soldier's Pocket Book for Field Service*, 4th edition (London: Macmillan, 1882)

Wrottesley, Lt.-Col. the Hon. G., *Life and Correspondence of Field Marshal Sir John Burgoyne, Bart*, 2 vols. (London: Richard Bentley & Son, 1873)

Wrottesley, Capt. the Hon. G. (ed.), *The Military Opinions of General Sir John Fox Burgoyne, Bart.* (London: Richard Bentley, 1859)

Younghusband, Capts. G. J. and F. E., *The Relief of Chitral* (London: Macmillan, 1895)

Contemporary articles and papers

Hamley, E., 'The Defencelessness of London', *Nineteenth Century*, vol. 23, no. 135 (May 1888), pp. 633–40

H. F., 'Our Baptism', *Vedette*, no. 109 (1898), pp. 2–22

Kunhardt, Capt. H. G., 'Notes on the Suakin–Berber Railway', *RE Journal*, vol. 15, no. 176 (1885), pp. 155–6

Luard, Capt. C. E., 'Field Railways, and their General Application in War', *Journal of the Royal United Service Institution*, vol. 17 (1873), pp. 693–724

Mance, Capt. H. O., 'Armoured Trains', *Professional Papers*, 4th series, vol. 1, no. 4 (1905), pp. 3–47

'Military Use of Railways in India: Precis of Report of the Railway Transport Committee, India, 1876', *Professional Papers of the Corps of Royal Engineers*, Occasional Papers, 2 (London: Royal Engineers Institute, 1878), pp. 22–39

Miller, Maj., 'The Italian Campaign of 1859', *Journal of the Royal United Service Institution*, vol. 5 (1861), pp. 269–308

Molesworth, Sir G. L., 'Railway Construction', *Professional Papers of the Corps of Royal Engineers*, Occasional Papers, 21 (Chatham: Royal Engineers Institute, 1895), pp. 137–256

Nathan, Lt. M., 'The Sudan Military Railway', *Professional Papers of the Corps of Royal Engineers*, vol. 11 (1885), pp. 35–49

Rothwell, Col. J. S., 'The Conveyance of Troops by Railway', *United Service Magazine*, vol. 4, no. 757 (1891), pp. 213–21

Satterthwaite, Lt. C. R., 'Some Notes on the Action of the Japanese First Army at Liao-Yang', *RE Journal*, vol. 17 (1913), pp. 343–54

Scott-Moncrieff, Capt. G. K., 'The Frontier Railways of India', *Professional Papers of the Corps of Royal Engineers*, vol. 11 (1885), pp. 213–56

——, 'Sir James Browne and the Harnae Railway', *Blackwood's Edinburgh Magazine*, vol. 177 (1905), pp. 608–21

——, 'Some Famous Engineer Officers of the Nineteenth Century', *RE Journal*, vol. 35 (1922), pp. 113–28

Simmons, Sir J. L. A., 'The Channel Tunnel', *Nineteenth Century*, vol. 11 (1882), pp. 663–7

'The Sind Pishin Railways', *Engineering*, vol. 14 (1888), pp. 364, 368–9, 381–4, 503–6

'Some Experiences with an Armoured Train in South Africa', *Highland Light Infantry Chronicle*, vol. 3, no. 6 (1902), pp. 743–7

Tyler, Capt. H. W., 'Railways Strategically Considered', *Journal of the Royal United Service Institution*, vol. 8 (1865), pp. 321–43

Wallace, Maj. W. A. J., 'Report of the Railway Operations in Egypt during August and September 1882', *Professional Papers of the Corps of Royal Engineers*, vol. 9 (1883), pp. 79–98

Wilkins, Lt.-Col. Henry St Clair, 'Abyssinian Expedition', report to Capt. T. S. Holland, *Papers on Subjects Connected with the Duties of the Corps of Royal Engineers*, new series, vol. 17 (1869), pp. 140–8

Willans, Lt., 'The Abyssinian Railway', *Papers on subjects Connected with the Duties of the Corps of Royal Engineers*, new series, vol. 18 (1870), pp. 163–76

Wing, Col. F. D. V., 'The Distribution and Supply of Ammunition on the Battlefield', *Journal of the Royal United Service Institution*, vol. 52 (1908), pp. 895–924

Wolseley, Gen. Viscount, 'The Standing Army of Great Britain', *Harper's New Monthly Magazine*, European edition, vol. 80 (1890), pp. 331–47

Wood, Lt.-Col. C. K., 'The Work of the Royal Engineers in Natal', *Professional Papers of the Corps of Royal Engineers*, Occasional Papers, 27 (Chatham: Royal Engineers Institute, 1901), pp. 49–70

Newspapers

Aberdeen Free Press
Argyllshire Herald
Ayr Advertiser
Birmingham Daily Gazette
Birmingham Daily Post
Bradford Observer; and Halifax, Huddersfield and Keighley Reporter

Brechin Advertiser
Bristol Observer
Bristol Times and Mirror
Cheltenham Chronicle
Coventry Herald
Crewe Guardian
Devon and Somerset Weekly News
Devon Weekly Times
[Doncaster] Chronicle
Dorset County Chronicle
Edinburgh Evening News
Fife News
Glasgow Herald
Gloucester Journal
Gloucestershire Chronicle
[Gloucestershire] Echo
Haddingtonshire Advertiser
Hampshire Observer
Horncastle News and South Lindsey Advertiser
Illustrated London News
Inverness Courier
John Bull
Leeds Mercury
Leicester Chronicle
Liverpool Mercury
[London] Daily News
Manchester Guardian
Manchester Evening News
Morning Leader
Nairnshire Telegraph
Newcastle Daily Chronicle
North Devon Herald
Northern Scot and Moray & Nairn Express
Northern Whig
North Star and Farmer's Chronicle
Pall Mall Gazette
Ross-shire Journal
Sheffield and Rotherham Independent
Sheffield Daily Telegraph
Somerset County Gazette
Stroud Journal
Tamworth Herald
The Times
Totnes Times
Warwick and Warwickshire Advertiser & Leamington Gazette
Western Morning News

SELECT BIBLIOGRAPHY

Secondary sources

Books

Alford, H. S. L. and W. Dennistoun Sword, *The Egyptian Soudan: Its Loss and Recovery* (London, 1898, reprinted London: Naval and Military Press, 1992)

Amery, L. S. (ed.), *The Times History of the War in South Africa 1899–1902*, 7 vols. (London: Sampson, Low, Marston & Co., 1900–9)

Arthur, Sir G., *Life of Lord Kitchener*, 3 vols. (London: Macmillan & Co., 1920)

Atkinson, C. T., *History of the Royal Dragoons 1661–1934* (Glasgow: R. Maclehose & Co., 1934)

Babington, A., *Military Intervention in Britain from the Gordon Riots to the Gibraltar Incident* (London: Routledge, 1991)

Bagwell, P. S., *The Transport Revolution from 1770* (London: Batsford, 1974)

Baring Pemberton, W., *Battles of the Boer War* (London: Pan Books, 1964)

Barthorp, M., *Afghan Wars and the North-West Frontier 1839–1947* (London: Cassell, 1982)

Bartlett, T., and K. Jeffery (eds.), *A Military History of Ireland* (Cambridge: Cambridge University Press, 1996)

Bates, D., *The Abyssinian Difficulty: The Emperor Theodorus and the Magdala Campaign 1867–68* (Oxford: Oxford University Press, 1979)

Beckett, I. F. W., *Riflemen Form: A Study of the Rifle Volunteer Movement, 1859–1908* (Aldershot: Ogilby Trusts, 1982)

——, *The Amateur Military Tradition 1558–1945* (Manchester: Manchester University Press, 1991)

——, *Victorians at War* (London: Hambledon, 2003)

—— (ed.), *Citizen Soldiers and the British Empire, 1837–1902* (London: Pickering & Chatto, 2012)

Berridge, P. S. A., *Couplings to the Khyber: The Story of the North Western Railway* (Newton Abbott: David & Charles, 1969)

Black, R. C. III, *The Railroads of the Confederacy* (Chapel Hill: University of North Carolina Press, 1952)

Bonavia, M. R., *The Organisation of British Railways* (London: Ian Allan, 1971)

Bond, B., *The Victorian Army and the Staff College, 1854–1914* (London: Eyre Methuen, 1972)

—— (ed.), *Victorian Military Campaigns* (London: Hutchinson, 1967)

Boyden, P. A., A. J. Guy and M. Harding (eds.), *Ashes and Blood: The British Army in South Africa, 1795–1914* (London: National Army Museum, 2001)

Callwell, Col. C. E., *Small Wars: A Tactical Textbook for Imperial Soldiers* (London: HMSO, 1896, reprinted London: Greenhill Books, 1990)

Cammack, D., *The Rand at War, 1899–1902: The Witwatersrand and the Anglo-Boer War* (London: James Currey, 1990)

Chaloner, W. H., *The Social and Economic Development of Crewe, 1780–1923* (Manchester: Manchester University Press, 1950)

Chandler, D., and I. Beckett (eds.) *The Oxford Illustrated History of the British Army* (Oxford: Oxford University Press, 1994)

Changuion, L., *Silence of the Guns: The History of the Long Toms of the Anglo-Boer War* (Pretoria: Protea Book House, 2001)

Charlton, J., *The Chartists: The First Nationalist Workers' Movement* (London: Pluto Press, 1997)

Chase, M., *Chartism: A New History* (Manchester: Manchester University Press, 2007)

Clark, J. E., Jr, *Railroads in the Civil War* (Baton Rouge: Louisiana State University Press, 2001)

Clarke, I. F., *Voices Prophesying War 1763–1984* (Oxford: Oxford University Press, 1966)

Cooke, B., *The Grand Crimean Central Railway*, 2nd revised edition (Knutsford: Cavalier House, 1997)

Creveld, M. van, *Supplying War: Logistics from Wallenstein to Patton* (Cambridge: Cambridge University Press, 1977)

Cunningham, H., *The Volunteer Force: A Social and Political History 1859–1908* (London: Croom Helm, 1975)

Daly, M. W., *The Sirdar: Sir Reginald Wingate and the British Empire in the Middle East* (Philadelphia: American Philosophical Society, 1997)

Darvall, F. O., *Popular Disturbances and Public Order in Regency England* (London: Oxford University Press, 1969)

David, S., *The Indian Mutiny* (London: Viking, 2002)

Dennis, P., and J. Grey (eds.), *The Boer War: Army, Nation and Empire. The 1999 Chief of Army/Australian War Memorial History Conference* (Canberra: Army History Unit, 2000)

Dunn-Pattison, R. P. *The History of the 91st Argyllshire Highlanders* (Edinburgh: Blackwood and Sons, 1910)

Edwardes, M., *Red Year: The Indian Rebellion of 1857* (London: Cardinal, 1975)

Emsley, C., *The English Police: A Political and Social History*, 2nd edition (Harlow, Essex: Longmans, 1996)

Fairrie, Lt.-Col. A., *'Cuidich N Righ': A History of the Queen's Own Highlanders (Seaforth and Camerons)* (Inverness: Regimental Headquarters, 1983)

Fergusson, T. G., *British Military Intelligence, 1870–1914: The Development of a Modern Intelligence Organization* (Frederick, Maryland: University Publications of America, Inc., 1984)

Figes, O., *Crimea: The Last Crusade* (London: Allen Lane, 2010)

Fleming, P., *The Siege at Peking* (Edinburgh: Birlinn, 1959)

Fraser-Tytler, Sir W. K., *Afghanistan: A Study of Political Developments in Central and Southern Asia*, 3rd revised edition (London: Oxford University Press, 1967)

Freeman, M., and D. Aldcroft, *The Atlas of British Railway History* (London: Croom Helm, 1985)

Friedberg, A. L., *The Weary Titan: Britain and the Experience of Relative Decline 1885–1905* (Princeton: Princeton University Press, 1988)

Gaylor, J., *Sons of John Company: The Indian and Pakistan Armies 1903–91*, 2nd edition (Tunbridge Wells, Kent: Parapress Ltd., 1996)

Gooch, J., *The Plans of War: The General Staff and British Military Strategy c. 1900–1916* (London: Routledge & Kegan Paul, 1974)

——, *The Prospect of War: Studies in British Defence Policy, 1847–1942* (London: Frank Cass, 1981)

—— (ed.), *The Boer War: Direction, Experience and Image* (London: Frank Cass, 2000)

Gourvish, T. R., *Mark Huish and the London and North Western Railway: A Study of Management* (Leicester: Leicester University Press, 1972)

——, *Railways and the British Economy 1830–1914* (Manchester: Manchester University Press, 1980)

Greaves, R. L., *Persia and the Defence of India 1884–1892: A Study in the Foreign Policy of the Third Marquis of Salisbury* (London: Athlone Press, 1959)

Greenhill Gardyne, Lt.-Col. G., *The Life of a Regiment: The History of the Gordon Highlanders from 1816 to 1898*, vol. 2 (London: The Medici Society, 1903)

Grundlingh, A., *The Dynamics of Treason: Boer Collaboration in the South African War of 1899–1902* (Pretoria: Protea, 2006)

Guy, A. J. (ed.), *The Road to Waterloo: The British Army and the Struggle against Revolutionary and Napoleonic France, 1793–1815* (Stroud, Gloucestershire: Alan Sutton for the National Army Museum, 1990)

Hay, D., and F. Snyder (eds.), *Policing and Prosecution in Britain, 1750–1850* (Oxford: Clarendon Press, 1989)

Haycock, R., and K. Neilson (eds.), *Men, Machines & War* (Waterloo, Ontario: Wilfrid Laurier University Press, 1988)

Heathcote, T. A., *The Military in British India: The Development of British Land Forces in South Asia 1600–1947* (Manchester: Manchester University Press, 1995)

Helps, A. *Life and Labours of Mr. Brassey* (London: Bell and Daldy, 1872)

Hibbert, C., *The Great Mutiny: India 1857* (London: Allen Lane, 1978)

Howard, M., *The Franco-Prussian War*, 2nd edition (London: Routledge, 2002)

Howie, D., *History of the 1st Lanark Volunteers* (Glasgow: David Robertson, 1887)

Huddleston, G., *History of the East Indian Railway* (Calcutta: Thacker, Spink and Co., 1906)

Hunter, A., *Kitchener's Sword Arm: The Life and Campaigns of General Sir Archibald Hunter* (Staplehurst: Spellmount, 1996)

Innes, Gen. J. J. McLeod, *The Life and Times of General Sir James Browne, R.E., K.C.B, K.C.S.I. (Buster Browne)* (London: John Murray, 1905)

Jackman, W. T., *The Development of Transportation in Modern England*, 2 vols. (Cambridge: Cambridge University Press, 1916)

Jeffery, K., *Field Marshal Sir Henry Wilson: A Political Soldier* (Oxford: Oxford University Press, 2006)

Jones, D. J. V., *The Last Rising: The Newport Insurrection of 1839* (Oxford: Clarendon Press, 1985)

——, *Rebecca's Children: A Study of Rural Society, Crime, and Protest* (Oxford: Clarendon Press, 1989)

Judd, D., and K. Surridge, *The Boer War* (London: John Murray, 2002)

Kazemzedah, F., *Russia and Britain in Persia, 1864–1914* (New Haven, Connecticut: Yale University Press, 1968)

Keown-Boyd, H., *A Good Dusting: A Centenary Review of the Sudan Campaigns 1883–1899* (London: Leo Cooper, 1986)

Kerr, I. J., *Building the Railways of the Raj* (Delhi: Oxford University Press, 1997)

Kestell, J. D., and D. E. van Velden, *The Peace Negotiations* (London: Richard Clay & Sons, 1912)

Kingsford, P. W., *Victorian Railwaymen: The Emergence and Growth of Railway Labour 1830–1870* (London: Cass, 1970)

Kochanski, H., *Sir Garnet Wolseley: Victorian Hero* (London: Hambledon Press, 1999)

Lewin, H. G., *The Railway Revolution*, volume 1: *Early British Railways*, reprinted with an introduction by Mark Casson (London: Routledge/Thoemmes Press, 1998)

Llewellyn-Jones, R., *The Great Uprising in India, 1857–58: Untold Stories, Indian and British* (Woodbridge: The Boydell Press, 2007)

Longford, E., *Wellington: Pillar of State* (London: Panther Books, 1975)

Luvaas, J., *The Military Legacy of the Civil War: The European Inheritance* (Chicago: University of Kansas Press, 1959)

——, *The Education of an Army: British Military Thought, 1815–1940* (London: Cassell, 1964)

McCracken, D. P., *MacBride's Brigade: Irish Commandos in the Anglo-Boer War* (Dublin: Four Courts Press, 1999)

Magnus, P., *Kitchener: Portrait of an Imperialist* (London: John Murray, 1958)

Mahajan, S., *British Foreign Policy, 1874–1914: The Role of India* (London: Routledge, 2002)

Marsh, P., *Beatty's Railway* (Oxford: New Cherwell Press, 2000)

Marshall, P. J. (ed.), *The Cambridge Illustrated History of the British Empire* (Cambridge: Cambridge University Press, 1996)

Mason, P., *A Matter of Honour: An Account of the Indian Army, its Officers and Men* (London; Jonathan Cape, 1974)

Mather, F. C., *Public Order in the Age of the Chartists* (Manchester: Manchester University Press, 1959)

Matthew, H. C. G., *Gladstone, 1875–1898* (Oxford: Clarendon Press, 1995)

Maxwell, L., *My God! – Maiwand* (London: Leo Cooper, 1979)

Miller, C., *Painting the Map Red: Canada and the South African War, 1899–1902* (Montreal and Kingston: Canadian War Museum and McGill-Queen's University Press, 1993)

Miller, S. M. (ed.), *Soldiers and Settlers in Africa, 1850–1918* (Leiden: Brill, 2009)

Morris, D. R., *The Washing of the Spears* (London: Sphere Books, 1968)

Mould, R. F., *A Century of X-Rays and Radioactivity in Medicine: With Emphasis on Photographic Records of the Early Years* (Bristol: Institute of Physics, 1993)

Neilson, K., *Britain and the Last Tsar: British Policy and Russia 1894–1917* (Oxford: Clarendon Press, 1995)

Nicoll, F., *Gladstone, Gordon and the Sudan Wars: The Battle over Imperial Intervention in the Victorian Age* (Barnsley: Pen & Sword, 2013)

Omissi, D. E., *The Sepoy and the Raj: The Indian Army 1860–1940* (Basingstoke: Macmillan, 1994)

Pakenham, T., *The Boer War* (London: Weidenfeld & Nicolson, 1979)

Palmer, S. H., *Police and Protest in England and Ireland 1780–1850* (Cambridge: Cambridge University Press, 1988)

Partridge, M. S., *Military Planning for the Defense of the United Kingdom, 1814–1870* (New York: Greenwood Press, 1989)

Plumridge, Lt.-Col. J. H., *Hospital Ships and Ambulance Trains* (London: Seeley, Service & Co., 1975)

Pollock, J., *Kitchener: The Road to Omdurman* (London: Constable, 1998)

Porch, D., *Wars of Empire* (London: Cassell, 2000)

Porter, A. (ed.), *The Oxford History of the British Empire*, vol. 3: *The Nineteenth Century* (Oxford: Oxford University Press, 1999)

Porter, W. Maj.-Gen., *History of the Corps of Royal Engineers*, vol. 2 (Chatham: Institution of Royal Engineers, 1889, reprinted 1977)

Powell, G., *Buller: A Scapegoat?* (London: Leo Cooper, 1994)

Pratt, E. A., *The Rise of Rail-Power in War and Conquest 1833–1914* (London: P. S. King & Son, 1915)

Preston, A., and P. Dennis, *Swords and Covenants* (London: Croom Helm, 1976)

Preston, D., *The Boxer Rebellion: The Dramatic Story of China's War on Foreigners that Shook the World in the Summer of 1900* (New York: Berkley Books, 1999)

Pretorius, F., *Historical Dictionary of the Anglo-Boer War* (Lanham, Maryland: Scarecrow Press, 2009)

Quinault, R., and J. Stevenson (eds.), *Popular Protest and Public Order: Six Studies in British History* (London: George Allen & Unwin, 1974)

Reid, R., *The Peterloo Massacre* (London: Heinemann, 1989)

Repington, C. à Court, *Vestigia: Reminiscences of Peace and War* (London: Constable, 1919)

Richards, J., and J. M. MacKenzie, *The Railway Station: A Social History* (Oxford: Oxford University Press, 1986)

Robbins, M., *George & Robert Stephenson* (London: Her Majesty's Stationery Office, 1981)

Roberts, B., *Kimberley: Turbulent City* (Cape Town: David Philip, 1976)

Robson, B., *Fuzzy-Wuzzy: The Campaigns in the Eastern Sudan 1884–85* (Tunbridge Wells: Spellmount, 1993)

——, *The Road to Kabul: The Second Afghan War 1878–1881* (London: Arms & Armour Press, 1996)

Rolt, L. T. C., *George and Robert Stephenson: The Railway Revolution* (London: Longmans, 1960)

Sandes, Lt.-Col. E. W. C., *The Military Engineer in India*, 2 vols. (Chatham: Institution of Royal Engineers, 1933)

——, *The Royal Engineers in Egypt and the Sudan* (Chatham: Institution of Royal Engineers, 1937)

——, *The Indian Sappers and Miners* (Chatham: Institution of Royal Engineers, 1948)

Satow, M., and R. Desmond, *Railways of the Raj* (London: Scolar Press, 1980)

Saville, J., *1848: The British State and the Chartist Movement* (Cambridge: Cambridge University Press, 1987)

Schivelbusch, W., *The Railway Journey: The Industrialisation of Time and Space in the Nineteenth Century* (Leamington Spa: Berg, 1986)

Scott Moncrieff, Maj.-Gen. Sir G., *Canals and Campaigns: An Engineer Officer in India 1877–1885* (London: British Association for Cemeteries in South Asia (BACSA), 1987)

Sheehan, W., and M. Cronin (eds.), *Riotous Assemblies: Rebels, Riots and Revolts in Ireland* (Cork: Mercier Press, 2011)

Simmons, J., *The Victorian Railway* (London: Thames and Hudson, 1991)

Simmons, J., and G. Biddle (eds.), *The Oxford Companion to British Railway History from 1603 to the 1990s* (Oxford: Oxford University Press, 1997)

Spiers, E. M., *The Army and Society, 1815–1914* (London: Longman, 1980)

——, *The Late Victorian Army 1868–1902* (Manchester: Manchester University Press, 1992)

——, *Wars of Intervention: A Case Study – The Reconquest of the Sudan 1896–99*, The Occasional, 32 (Camberley: Strategic & Combat Studies Institute, 1998)

—— (ed.), *Sudan: The Reconquest Reappraised* (London: Frank Cass, 1998)

——, J. A. Crang and M. J. Strickland (eds.), *A Military History of Scotland* (Edinburgh: Edinburgh University Press, 2012)

Spies, S. B., *Methods of Barbarism? Roberts and Kitchener and Civilians in the Boer Republics, January 1900 – May 1902* (Cape Town: Human & Rousseau, 1977)

Stanmore, Lord, *Sidney Herbert: Lord Herbert of Lea: A Memoir*, 2 vols. (London: John Murray, 1906)

Stevenson, J., *Popular Disturbances in England 1700–1832*, 2nd revised edition (London: Longmans, 1992)

Stover, J. F., *American Railroads*, 2nd edition (Chicago: University of Chicago Press, 1997)

Strachan, H., *Wellington's Legacy: The Reform of the British Army, 1830–54* (Manchester: Manchester University Press, 1984)

——, *The Politics of the British Army* (Oxford: Clarendon Press, 1997)

—— (ed.), *Big Wars and Small Wars: The British Army and the Lessons of War in the Twentieth Century* (London: Routledge, 2006)

Symons, J., *England's Pride: The Story of the Gordon Relief Expedition* (London: Hamish Hamilton, 1965)

Theobald A. B., *The Mahdiya: A History of the Anglo-Egyptian Sudan, 1881–1899* (London: Longmans Green, 1951)

Townsend, Maj. C. E. C., *All Rank and No File: A History of the Engineer and Railway Staff Corps, RE, 1865–1965* (London: Engineer & Railway Staff Corps RE (TAVR), 1965)

Turner, G. E., *Victory Rode the Rails* (New York: Bobbs-Merrill, 1953)

Walmsley, R., *Peterloo: The Case Reopened* (Manchester: Manchester University Press, 1969)

Watson, Col. Sir C. M., *History of the Corps of Royal Engineers* (Chatham: Institution of Royal Engineers, 1934)

Weber, T., *The Northern Railroads in the Civil War* (New York: King's Crown, 1952)

Welsby, D. A., *Sudan's First Railway: The Gordon Relief Expedition and the Dongola Campaign*, Sudan Archaeological Research Society Publication 19 (London: Sudan Archaeological Research Society, 2011)

Westwood, J. N., *Railways of India* (Newton Abbot: David & Charles, 1974)

——, *Railways at War* (London: Osprey, 1980)

White-Spurner, B., *Horse Guards* (London: Macmillan, 2006)

Williams, C. (ed.), *A Companion to Nineteenth-Century Britain* (Oxford: Blackwell, 2004)

Williams, D., *The Rebecca Riots: A Study in Agrarian Discontent* (Cardiff: University of Wales Press, 1955)

Williams, G. A., *The Merthyr Rising* (London: Croom Helm, 1978)

Williams, G. F., *The Diamond Mines of South Africa: Some Account of their Rise and Development* (New York: Macmillan, 1902)

Williamson, S. R., Jr., *The Politics of Grand Strategy: Britain and France Prepare for War, 1904–1914* (Cambridge, Massachusetts: Harvard University Press, 1969)

Wilson, K. M., *Channel Tunnel Visions 1850–1945: Dreams and Nightmares* (London: Hambledon Press, 1994)

Wingate, F. R., *Mahdism and the Egyptian Sudan*, 2nd edition (London: Frank Cass, 1968)

Wolmar, C., *Engines of War: How Wars were Won and Lost on the Railways* (London: Atlantic, 2010)

Ziegler, P., *Omdurman* (London: Collins, 1973)

Zulfo, I. H., *Karari: The Sudanese Account of the Battle of Omdurman*, trans. P. Clark (London: Frederick Warne, 1980)

Articles

Anderson, Lt.-Col. R. B., 'The Nile Circus: An Uncomfortable Entertainment in which the Scots Greys got the Hump', *Eagle and Carbine* (1985), pp. 181–96

Bailes, H., 'Technology and Imperialism: A Case Study of the Victorian Army in Africa', *Victorian Studies*, vol. 24 (1980), pp. 83–104

Barrie, D. G., 'A Typology of British Police: Locating the Scottish Municipal Police Model in its British Context, 1800–1835', *British Journal of Criminology*, vol. 50 (2010), pp. 259–77.

Beckett, I. F. W., 'The Stanhope Memorandum of 1888: A Reinterpretation', *Bulletin of the Institute of Historical Research*, vol. 57, no. 136 (1984), pp. 240–7

——, 'Soldiers, the Frontier and the Politics of Command in British India', *Small Wars and Insurgencies*, vol. 16 (2005), pp. 280–92

Buttery, D., 'A "Toff's Life" in the Blockhouse', *Soldiers of the Queen*, no. 103 (2000), pp. 21–3

Donajgrodzki, A. P., 'Sir James Graham at the Home Office', *Historical Journal*, vol. 20, no. 1 (1977), pp. 97–120

Hart, J. M., 'The Reform of the Borough Police, 1835–56', *English Historical Review*, vol. 70 (1955), pp. 411–27

Hill, R., 'The Suakin–Berber Railway, 1885', *Sudan Notes and Records*, vol. 20, no. 1 (1937), pp. 107–24

H.L.P., 'Colonel Sir Percy Edouard Cranwill Girouard', *RE Journal*, vol. 47, no. 2 (1933), pp. 323–43

Johnson, R. A., 'The Penjdeh Incident, 1885', *Archives*, vol. 24 (1999), pp. 28–48

——, '"Russians at the Gates of India"? Planning the Defense of India, 1885–1900', *Journal of Military History*, vol. 67, no. 3 (2003), pp. 697–743

Jones, D. J. V., 'The Merthyr Riots of 1831', *Welsh History Review*, vol. 3, no. 2 (1966), pp. 173–205

——, 'Law Enforcement and Popular Disturbances in Wales, 1793–1835', *Journal of Modern History*, vol. 42 (1970), pp. 496–523

Lera, N., 'The Baluchistan "White Elephant": The Chappar Rift and Other Strategic Railways on the Border of British India', *Asian Affairs*, vol. 31, no. 2 (2000), pp. 170–80

Mains, Lt.-Col. A. A., 'The Auxiliary Force (India)', *Journal of the Society for Army Historical Research*, vol. 61, no. 247 (1983), pp. 160–85

Martin, B., 'The Opening of the Railway between Durban and Pietermaritzburg – 100 Years Ago', *Natalia*, vol. 10 (1980), pp. 35–40

Mather, F. C., 'The Railways, the Electric Telegraph and Public Order during the Chartist Period, 1837–48', *History*, vol. 38 (1953), pp. 40–53

Nasson, B., 'Moving Lord Kitchener: Black Military Transport and Supply Work in the South African War, 1899–1902, with Particular Reference to the Cape Colony', *Journal of Southern African Studies*, vol. 11, no. 1 (1984), pp. 25–51

Neilson, K., '"That Dangerous and Difficult Enterprise": British Military Thinking and the Russo-Japanese War', *War & Society*, vol. 9, no. 2 (1991), pp. 17–37

Paley, R., '"An imperfect, inadequate and wretched system?" Policing London before Peel', *Criminal Justice History*, vol. 10 (1989), pp. 95–130

Poole, R., 'By the Law or the Sword: Peterloo Revisited', *History*, vol. 91 (2006), pp. 254–76

Radcliffe, General Sir P., '"With France": The "WF" Plan & the Genesis of the Western Front', *Stand To!*, vol. 11 (1984), pp. 6–12

Robbins, M., 'The Balaklava Railway', *Journal of Transport History*, vol. 1, no. 1 (1953), pp. 28–43

Schölch, A., 'The "Men on the Spot" and the English Occupation of Egypt in 1882', *Historical Journal*, vol. 19 (1976), pp. 773–85

'Some Recollections of the Zulu War, 1879: Extracted from the Unpublished Reminiscences of the Late Lieut.-General Sir Edward Hutton, KCB, KCMG', *Army Quarterly*, vol. 26 (1928), pp. 65–80

Spiers, E. M., 'Intelligence and Command in British Small Colonial Wars of the 1990s', *Intelligence and National Security*, vol. 22, no. 5 (2007), pp. 661–81

——, 'The Learning Curve in the South African War: Soldiers' Perspectives', *Historia*, vol. 55, no. 1 (2010), pp. 1–17

Teichman, Maj. O., 'The Yeomanry as an Aid to Civil Power, 1795–1867', *Journal of the Society for Army Historical Research*, vol. 19 (1940), pp. 75–91 and 127–43

Ward, S. G. P., 'The Scots Guards in Egypt, 1882', *Journal of the Society for Army Historical Research*, vol. 51 (1973), pp. 80–104

Zurnamer, Maj. B. A., 'The State of the Railways in South Africa during the Anglo-Boer War 1899–1902', *Scientia Militaria: South African Journal of Military Studies*, vol. 16, no. 4 (1986), pp. 26–33, http://scientiamilitaria.journals.ac.za (accessed 18 April 2014)

INDEX

INDEX

INDEX

INDEX

INDEX